. . . ensure adequate protection of public health and safety, promote the common defense and security, and protect the environment.

U.S. NUCLEAR REGULATORY COMMISSION

FY 2005 PERFORMANCE AND ACCOUNTABILITY REPORT

MISSION

License and regulate the Nation's civilian use of byproduct, source, and special nuclear materials to ensure adequate protection of public health and safety, promote the common defense and security, and protect the environment.

VISION

Excellence in regulating the safe and secure use and management of radioactive materials for the public good.

TABLE OF CONTENTS

(From left to right) Commissioner Gregory B. Jaczko, Commissioner Edward McGaffigan, Jr., Chairman Nils J. Diaz, Commissioner Jeffrey S. Merrifield, and Commissioner Peter B. Lyons

A MESSAGE FROM THE CHAIRMAN

I am pleased to present the Nuclear Regulatory Commission's *Performance and Accountability Report for FY 2005*. The NRC's strategic objective is to enable the use and management of radioactive material and nuclear fuels for beneficial civilian purposes in a manner that protects public health and safety and the environment, promotes the security of our Nation, and provides for regulatory actions that are open, effective, efficient, realistic, and timely. I am proud to report that once again the NRC has achieved all of its safety and security performance goals. I am also pleased to report that the NRC continues to position its resources and infrastructure to maintain its strong oversight of existing facilities and to review future applications for new power reactors, license renewals for existing facilities, and a high-level waste repository.

Safety and security continues to be NRC's highest priority. The NRC has strengthened safety and security at commercial nuclear facilities and augmented protection of NRC-regulated radioactive material. In addition, we increased efforts to improve preparedness and emergency response capabilities significantly. As a result, America's nuclear facilities are more secure and better defended today than ever before.

The NRC is committed to ensuring that our resources are well managed. This report provides information which demonstrates the prudent management of the funds entrusted to us by the American public and describes our successes in implementing the President's Management Agenda to promote a more efficient and effective Government. The Reports Consolidation Act requires an assessment of the completeness and reliability of the program performance and financial data contained in this report based on evaluation criteria issued by the Office of Management and Budget. I conclude that the data are complete and reliable. In addition, the NRC has evaluated its management controls and financial management systems as required by the Federal Managers' Financial Integrity Act. On the basis of the NRC's comprehensive management control program, I certify, with reasonable assurance, that the NRC is in compliance with the provisions of this act. The auditors have issued an unqualified opinion on our FY 2005 financial statements. They have identified the quality assurance process over the Fee Billing System as a material internal control weakness. The auditors also identified the Fee Billing System, as well as the core accounting and payroll systems provided by the Department of Interior's National Business Center as being in substantial noncompliance with Governmentwide financial system requirements and with the Federal Financial Management Improvement Act. We have implemented improvements to the fee billing process (see Chapter 1, *Audit Results*) and will continue to work toward eliminating this material internal control weakness and Federal Financial Management Improvement Act substantial noncompliance.

The NRC's vision is to achieve excellence in regulating the safe and secure use and management of radioactive materials for the public good. The Commission is proud of this past fiscal year's accomplishments in fulfilling the vision and looks forward to continuing to provide high-quality service to the licensees we regulate and to the American public.

Nils J. Diaz
November 15, 2005

. . . NRC's
best asset is
its staff . . .

INTRODUCTION

This *Performance and Accountability Report* represents the culmination of the U.S. Nuclear Regulatory Commission's (NRC) program and financial management processes, which began with mission and program planning, continued through the formulation and justification of NRC's budget to the President and the Congress, through budget execution, and ended with this report on our program performance and use of the resources entrusted to us. This report was prepared pursuant to the requirements of the Chief Financial Officers Act, as amended by the Reports Consolidation Act, and covers activities from October 1, 2004, to September 30, 2005. This report can be accessed on NRC's Web site at http://www.nrc.gov.

Chapter 1, *Management's Discussion and Analysis*, provides an overview of the NRC. It consists of six sections: *About the NRC* describes the agency's mission, organizational structure, and regulatory responsibility; *Future Challenges* includes forward-looking information; *Program Performance Overview* discusses the agency's success in achieving its strategic goals which are further described in Chapter 2; *President's Management Agenda* describes the agency's progress in "Getting to Green" for the five management initiatives; *Financial Performance Overview* provides highlights of the NRC's financial position and audit results contained in Chapter 3; and *Systems, Controls, and Legal Compliance* describes the agency's compliance with key legal and regulatory requirements.

ABOUT THE NRC

The NRC was established on January 19, 1975, as an independent Federal agency to regulate commercial and institutional uses of nuclear materials. The NRC's purpose is defined by the Atomic Energy Act, as amended, and the Energy Reorganization Act, as amended. These acts provide the foundation for regulating the Nation's civilian uses of nuclear materials.

The NRC's mission is to regulate the Nation's civilian use of byproduct, source, and special nuclear materials to ensure adequate protection of public health and safety, to promote the common defense and security, and to protect the environment.

Organization

The NRC is headed by a Commission composed of five members, with one member designated by the President to serve as Chairman. Each member is appointed by the President, with the advice and consent of the Senate, to serve 5-year terms. The Chairman serves as the principal executive officer and official spokesman for the Commission. The Executive Director for Operations carries out the program policies and decisions made by the Commission.

The NRC's headquarters is located in Rockville, Maryland. Four regional offices are located in King of Prussia, Pennsylvania; Atlanta, Georgia; Lisle, Illinois; and Arlington, Texas. The NRC's technical training center is located in Chattanooga, Tennessee. The NRC also has at least two resident inspectors at each of the Nation's nuclear power reactor sites. The NRC's Operations Center is the focal point for NRC communications with its licensees, State agencies, and other Federal agencies concerning operating events in the commercial nuclear sector. The Operations Center is staffed 24 hours a day by NRC operations officers. An organization chart of the NRC is contained in Appendix E.

The NRC's budget for fiscal year (FY) 2005 was $669.3 million (see Figure 1) and 3,108 full-time equivalent staff (see Figure 2). The FY 2004 budget was $625.6 million and 3,040 full-time equivalent staff. The NRC recovers most of its appropriations from fees paid by NRC licensees. Approximately 66 percent of the budget and 69 percent of the staff are for reactor safety.

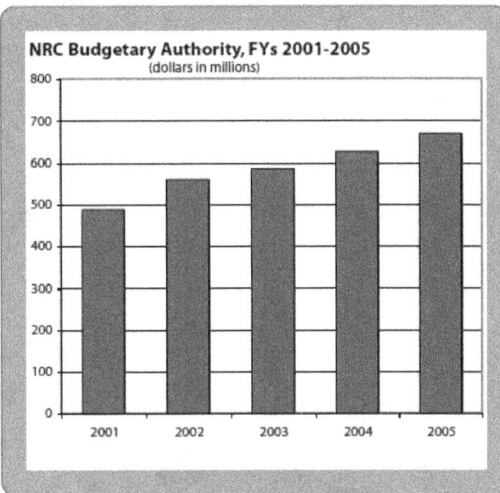

Figure 1

Regulatory Responsibility

To fulfill its responsibility to protect the public health and safety, the NRC performs three principal regulatory functions (1) establish standards and regulations, (2) issue licenses for nuclear facilities and users of nuclear materials, and (3) inspect facilities and users of nuclear materials to ensure compliance with regulatory requirements. These regulatory functions relate to nuclear power plants, other nuclear facilities, and other civilian uses of nuclear materials, such as nuclear medicine programs at hospitals; academic activities at educational institutions; research work; industrial applications, such as

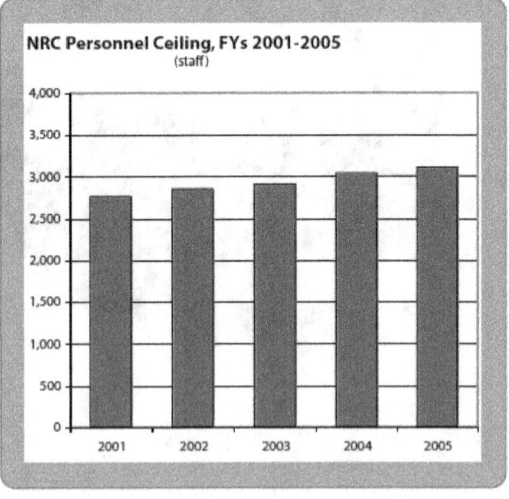

Figure 2

gauges and testing equipment; and the transport, storage, and disposal of nuclear materials and wastes. The NRC has aligned its regulatory activities into the Nuclear Reactor Safety program and the Nuclear Materials and Waste Safety program.

The NRC also carries out a corporate management and support function for information technology, financial management, human resources, administrative services, and other support functions. Efforts in this area are aligned with the President's Management Agenda and focus on the five Governmentwide initiatives aimed at improving agency management.

Approximately 20 percent of the Nation's electricity is generated by 104 commercial nuclear reactors that are licensed by the NRC to operate in 31 States (see Figure 3). Since 1993, nuclear electric generation has increased by approximately 20 percent. The NRC expends over 368,500 hours of inspection effort annually at 104 operating reactors and licenses approximately 4,700 reactor operators.

The NRC oversees approximately 4,500 licenses for medical, academic, industrial, and general uses of nuclear materials. The NRC conducts between 1,150 and 1,200 health and safety inspections of its nuclear materials licensees annually. Additionally, approximately 17,300 licenses are administered by the 33 States that participate in the NRC Agreement States program, which authorizes the State to regulate the use of radioactive materials within that State (see Figure 4). The NRC, Agreement States, and their licensees share a common responsibility to protect public health and safety.

The NRC places a high priority on keeping the public informed of its activities. Visit our Web site at http://www.nrc.gov to learn more about who we are and what we do to serve the American people.

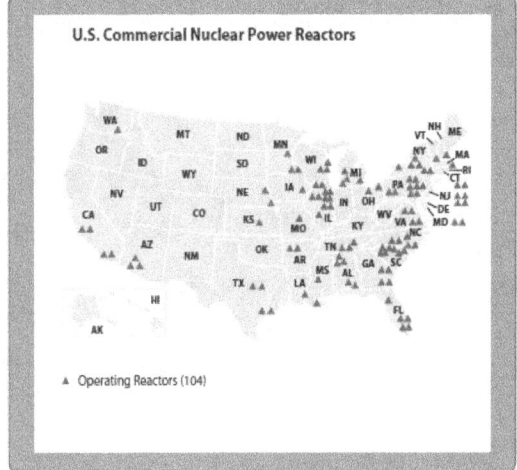

U.S. Commercial Nuclear Power Reactors

▲ Operating Reactors (104)

Figure 3

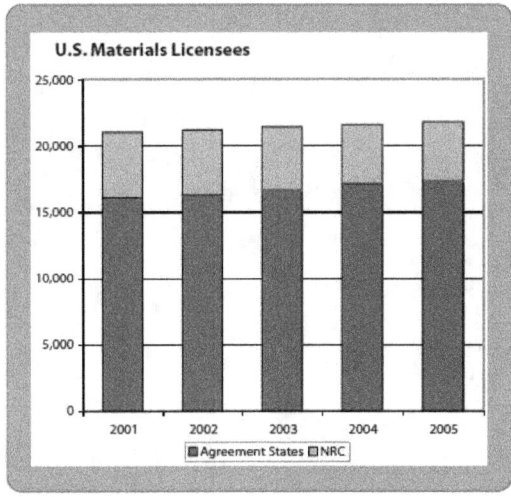

U.S. Materials Licensees

Figure 4

FUTURE CHALLENGES

The Commission is focused on addressing a number of significant challenges that will have a long-term impact in accomplishing its mission. The many industries that utilize radioactive materials are experiencing change, particularly in the areas of nuclear safety, security and emergency preparedness, risk-informed and performance-based regulations, energy production, and waste management. In the next 5 years, the Nation is likely to see the following changes occur:

- NRC strategic initiatives will include continued emphasis on strengthening the interrelationship between safety, security, and emergency preparedness.

- The NRC will continue to be challenged by the need to excel in the management of safety and in the area of communications. One of our greatest continuing challenges lies in the area of integrating power plant security, and its many improvements, into the fabric of day-to-day operational safety and regulation. We also need to continue to enhance integration of our security-related interfaces with other agencies, at the Federal, State, and local levels.

- The NRC must continue to foster and strengthen, internally and externally, the means of achieving safety-focused policy, programs, and practices, directed by and with the ultimate goal of benefiting the American public.

- The nuclear industry needs to continue to foster and strengthen design, operational and safety maintenance, making the right choices, and avoiding the pitfall of complacency or lack of safety focus.

- The majority of operating nuclear power plants will apply for license renewal to help meet the Nation's demand for energy production.

- The Department of Energy (DOE) is expected in the near future to submit an application to construct and operate the Nation's high-level radioactive waste repository at Yucca Mountain, Nevada.

- Increasing quantities of radioactive waste may be transported and held in interim storage or permanent disposal sites.

- The nuclear power industry will begin to submit applications to construct and operate new nuclear power plants to meet the Nation's demand for energy.

- The NRC, Agreement States, and licensees will continue to devote increasing attention to the security of radioactive materials and facilities; in addition, the NRC will continue its nuclear non-proliferation activities.

- The NRC will continue to see an increase in requirements for coordination with a wide array of Federal, State, and local agencies related to homeland security and emergency planning.

- The number of Agreement States will increase, as will the numbers of medical, academic, and industrial entities using radioactive materials under the oversight of the Agreement States.

- The regulatory climate is expected to adjust to both internal and external factors through the use of risk-informed and, as appropriate, performance-based regulations.

A backdrop to these industry-specific changes is one of elevated security and heightened public concern about safety. This has resulted in increased public dialogue about the uses of radioactive materials, varying from the potential for terrorist activities, to public concern about the adequacy of emergency preparedness plans for areas surrounding nuclear facilities. In this regard, the NRC is committed to sharing openly with the public its information and decision-making processes consistent with the law and is committed to implementing regulatory processes that facilitate stakeholder involvement. While the NRC will continue to make as much information as possible available to the public, the agency will withhold information that could assist potential terrorists. The manner in which the NRC regulates is also evolving. As the NRC continues to learn from operational experience and develops more effective ways of assessing risks and using risk-informed and performance-based approaches founded in 'realistic conservatism,' the agency is better able to make appropriate safety decisions and to better allocate resources to areas where they will have the greatest positive effect. In addition, the NRC continues to seek improvement in effectiveness and efficiency.

With respect to all facilities licensed by the NRC and Agreement States, the NRC is increasingly approaching safety, security, incident response, and emergency preparedness in an integrated manner. Safety requirements for structures, systems, components, programs, and people all contribute to both safety and security by making accidents unlikely and by making mitigation capability strong. In addition, safety and security requirements work together to make these facilities uninviting targets.

Ensuring the protection of public health, safety, and the environment has always been, and continues to be, the NRC's primary goal. Accordingly, safety is the most important consideration in evaluating license applications, licensee performance, and proposed changes to the regulatory framework. Because security is essential to the NRC's mission and linked to safety, it also is an important consideration in the agency's actions. The agency continuously works to improve its effectiveness and efficiency without conflicting with or undermining its safety and security mission.

PROGRAM PERFORMANCE OVERVIEW

Federal agencies provide an annual performance budget to Congress with measurable target levels of performance. The NRC evaluates its program performance within a structured planning, budgeting, and performance management process.

Strategic Plan

The NRC's FY 2004–FY 2009 Strategic Plan describes the strategies and means by which the agency intends to accomplish its mission. The Strategic Plan provides a foundation to guide the development of the NRC's annual performance plan and subsequent resource requirement determination. The Strategic Plan focuses on the agency's strategic objective to "enable the use and management of radioactive materials and nuclear fuels for beneficial civilian purposes in a manner that protects public health and safety and the environment, promotes the security of our Nation, and provides for regulatory actions that are open, effective, efficient, realistic, and timely." The Strategic Plan also contains the agency's five goals of safety, security, openness, effectiveness, and management. Safety and security are the agency's highest priorities and are supported by the goals of openness, effectiveness, and management. The agency's success in achieving each goal is defined by strategic outcome(s) that are associated with each goal, which is described below. A further description of each goal and their associated performance measures are provided in Chapter 2, *Program Performance*, which describes the agency's overall program performance.

Goals

Safety goal—*Ensure protection of public health and safety and the environment.*

Strategic Outcomes

- No nuclear reactor accidents.[1]

- No inadvertent criticality events.

- No acute radiation exposures resulting in fatalities.

- No releases of radioactive materials that result in significant radiation exposures.

- No releases of radioactive materials that cause significant adverse environmental impacts.

The NRC's primary goal is to regulate the safe uses of radioactive materials for civilian purposes to ensure the protection of public health and safety and the environment. The NRC achieves its safety goal by licensing individuals and organizations to use radioactive materials for beneficial civilian purposes and then ensuring that the performance of these licensees is at or above acceptable safety levels. In particular, we maintain vigilance over safety performance through ongoing licensing reviews and inspections and expanded oversight. We also use enforcement actions for significant deficiencies, including issuing orders for corrective action, issuing shutdown orders, imposing civil penalties and/or criminal prosecution, or, when appropriate, suspending or revoking a license.

Security goal—*Ensure the secure use and management of radioactive materials.*

Strategic Outcome

- No instances where licensed radioactive materials are used domestically in a manner hostile to the security of the United States.

The security goal has been explicitly identified to reflect changes in the threat environment and the agency's response to the events of September 11, 2001. The primary challenge facing the NRC in the coming years is to emerge from the period of uncertainty in post-September 11 security requirements; determine what long-term security provisions are necessary; and revise its regulations, orders, and internal procedures as necessary to ensure public health and safety and the common defense and security in an elevated threat environment. In particular, the NRC will focus its efforts on the following activities:

- Assure the continuing validity of the NRC design-basis threats.

- Complete the identification of vulnerabilities and mitigating strategies at licensed facilities.

- Revise requirements to provide additional protection where needed. Develop improved methods of communicating sensitive information to licensees.

- Enhance controls on high-risk radiation sources.

- Develop more formal, long-term relationships with Federal, State, and local organizations with shared responsibilities for protecting nuclear facilities and activities and responding to incidents.

<u>Openness goal</u>—*Ensure openness in our regulatory process.*

Strategic Outcome

- Stakeholders are informed and involved in NRC processes as appropriate.

The goal on openness recognizes that stakeholders be informed about, and have an opportunity to participate in the NRC's regulatory process. The NRC views nuclear regulation as the public's business; as such, it should be transacted openly and candidly in order to maintain the public's confidence. The agency is committed to keeping the public informed and believes that a responsible and effective regulatory process includes an involved public that is well informed.

<u>Effectiveness goal</u>—*Ensure that NRC actions are effective, efficient, realistic, and timely.*

Strategic Outcome

- No significant licensing or regulatory impediments to the safe and beneficial uses of radioactive materials.

Over the next several years, the NRC anticipates a significant increase in agency workload. In particular, the workload is likely to include licensing requests of unprecedented technical complexity, including the Department of Energy's application to license the Yucca Mountain high-level radioactive waste repository and requests to license the next generation of nuclear reactors. Security demands are becoming more complex, requiring diverse professional expertise and close coordination with other Federal, State, and local agencies. Increases in both the frequency and the extent of stakeholder involvement in the NRC's regulatory processes are expected as the agency works to improve openness.

Many factors could contribute to licensing and regulatory impediments, such as an inadequate regulatory framework, an ineffective program, or an inefficient process that results in an untimely regulatory decision. The NRC is committed to addressing such issues through initiatives related to this goal, and it will also monitor the regulated community for instances where agency actions may have unnecessarily impeded licensees and applicants. In conducting this monitoring, the NRC may consider the results of self-assessments and external assessments, feedback from stakeholders, congressional direction, and other sources.

<u>Management goal</u>—*Ensure excellence in agency management to carry out the NRC's strategic objective.*

Strategic Outcomes

- Continuous improvement in NRC's leadership and management effectiveness in delivering the mission.

- A diverse, skilled workforce and an infrastructure that fully support the agency's mission and goals.

The NRC strives for management excellence in carrying out all of its regulatory responsibilities. The agency believes that management excellence should be achieved while fostering the successful conduct of priority activities. In setting this goal, the NRC considered the management and support needed to achieve the agency's mission, preexisting management challenges, and other initiatives identified by central organizations such as the Government Accountability Office (GAO), Office of Management and Budget (OMB), and Office of Personnel Management (OPM). This goal includes strategies for the management of human capital, infrastructure management, improved financial performance, expanded electronic Government, budget and performance integration, and internal communications.

The NRC's FY 2004–FY 2009 Strategic Plan is available on the Web site at http://www.nrc.gov/reading-rm/doc-collections/nuregs/staff/sr1614/v3/index.html.

Performance Budget

The FY 2004–FY 2009 Strategic Plan led to a re-alignment of the agency's performance budget structure. Beginning with the FY 2006 Performance Budget, budget requests have been structured by two major programs—Nuclear Reactor Safety and Nuclear Materials and Waste Safety.

Nuclear Reactor Safety

The Nuclear Reactor Safety program encompasses all NRC efforts to ensure that civilian nuclear power reactor facilities and research and test reactors are licensed and operated in a manner that adequately protects the public health and safety and the environment and protects against radiological sabotage and theft or diversion of special nuclear materials. The Atomic Energy Act of 1954, as amended, and the Energy Reorganization

Act of 1974, as amended, are the foundation for regulating the Nation's civilian nuclear power industry. The Nuclear Reactor Safety program contains two activities—Nuclear Reactor Licensing and Nuclear Reactor Inspection.

Nuclear Materials and Waste Safety

The NRC protects the public health and safety and the environment and ensures the secure use and management of radioactive materials through the Nuclear Materials and Waste Safety program. The Nuclear Materials and Waste Safety program contains five activities—Fuel Facilities Licensing and Inspection, Nuclear Materials Users Licensing and Inspection, High-Level Waste Repository, Decommissioning and Low-Level Waste, and Spent Fuel Storage and Transportation Licensing and Inspection.

The NRC's FY 2006 Performance Budget is available on the Web site at http://www.nrc.gov/reading-rm/doc-collections/nuregs/staff/sr1100/v21/.

Program Performance Report

The *FY 2005 Performance and Accountability Report* is reporting on the agency's performance measures by each of the five goals contained in the agency's FY 2004–FY 2009 Strategic Plan. The *FY 2005 Performance and Accountability Report* also describes the achievements and challenges faced by each of the seven activities under the agency's major programs of Nuclear Reactor Safety and Nuclear Materials and Waste Safety. The NRC is reporting FY 2005 performance measures contained in both the FY 2005 and FY 2006 Performance Budgets.

Program Assessment Rating Tool

Over the past several years, the Office of Management and Budget has conducted Program Assessment Rating Tool reviews of the NRC's Nuclear Reactor Safety and the Nuclear Materials and Waste Safety program activities. All the program activities reviewed have been rated as either "effective" or "moderately effective." For the program activities reviewed in FY 2003 and FY 2004, the Office of Management and Budget recommended that the agency include better linkages of budget requests to the NRC's annual and long-term goals, as well as the linkage of performance measures in the organization's operating plan to support the safety performance measures in the agency's FY 2004–FY 2009 Strategic Plan. A further recommendation was for more transparency in how allocation decisions are made and how the activity contributes to achievement of the agency's long-term goals. In addition, the Office of Management and Budget recommended a complete review of operating plan format and content to improve their effectiveness as management tools. The NRC has been addressing the recommendations from the Office of Management and Budget to improve the effectiveness of these program activities.

PRESIDENT'S MANAGEMENT AGENDA

The President's Management Agenda prescribes Governmentwide initiatives to reform the U.S. Government to be more citizen-centered, results-oriented, and market-based, and to actively promote competition rather than stifling innovation. To achieve this goal, the Administration has identified five initiatives to improve Government performance in the areas of (1) strategic management of human capital, (2) budget and performance integration, (3) competitive sourcing, (4) expanded electronic Government, and (5) improved financial management. The NRC is actively implementing the agenda to improve the management and performance of the Federal Government. Chapter 2 of this report discusses our accomplishments in these important areas.

FINANCIAL PERFORMANCE OVERVIEW

As of September 30, 2005, and 2004, the financial condition of the NRC was sound with respect to having sufficient funds to meet program needs and adequate control of these funds in place to ensure obligations did not exceed budget authority. The NRC prepared its financial statements in accordance with the accounting standards codified in the Statements of Federal Financial Accounting Standards (SFFAS) and Office of Management and Budget Circular A-136, *Financial Reporting Requirements*.

Sources of Funds

The NRC has two appropriations, Salaries and Expenses and Office of the Inspector General, and funds for both appropriations are available until expended. The NRC's total new FY 2005 budget authority was $669.3 million. Of this amount, $661.8 million was for the Salaries and Expenses appropriation and $7.5 million was for the Office of the Inspector General appropriation. This represents an increase in new budget authority of $43.6 million over FY 2004 ($43.4 million for the Salaries and Expenses appropriation and $0.2 million for the Office of the Inspector General appropriation). In addition, $40.9 million from prior-year appropriations, $6.1 million from prior-year reimbursable work, and $6.6 million for new reimbursable work to be performed for others was available to obligate in FY 2005. The sum of all funds available to obligate for FY 2005 was $722.9 million, which is a $41.3 million increase over the FY 2004 amount of $681.6 million.

The Omnibus Budget Reconciliation Act of 1990 (OBRA-90), as amended, required the NRC to collect fees to offset approximately 90 percent of its new budget authority, less the amount appropriated to the NRC from the Nuclear Waste Fund for FY 2005 (see Figure 5). The NRC collected $534.1 million in FY 2005. This is 98.8 percent of the recovery requirement. For FY 2004, OBRA-90 required NRC to collect approximately 92 percent of its new budget authority, excluding appropriations from the Nuclear Waste Fund.

Uses of Funds by Function

The NRC incurred obligations of $665.5 million, which was an increase of $20.3 million over FY 2004. Approximately 60 percent of obligations were used for salaries and benefits. The remaining 40 percent was used to obtain technical assistance for the NRC's principal regulatory programs, to conduct confirmatory safety research, to cover operating expenses, (e.g., building rentals, transportation, printing, security services, supplies, office automation, training), staff travel, and reimbursable work (see Figure 6). The unobligated budget authority available at the end of FY 2005 was $57.3 million, which is an increase compared to the FY 2004 amount of $36.3 million. This increase in year-end unobligated budget authority is primarily the result of the delay in the Department of Energy's submission of a license application for a high-level waste repository for NRC review. Of this $57.3 million, $6.7 million is for reimbursable work and $50.6 million is available to fund critical NRC needs in FY 2006.

Audit Results

The NRC received an unqualified audit opinion on its FY 2005 financial statements. The auditors identified two new reportable conditions concerning information system security and financial controls over disbursements. In FY 2004, the auditors identified one reportable condition, which was classified as a material internal control weakness concerning the Fee Billing System. This weakness was also identified as a substantial noncompliance with the Federal

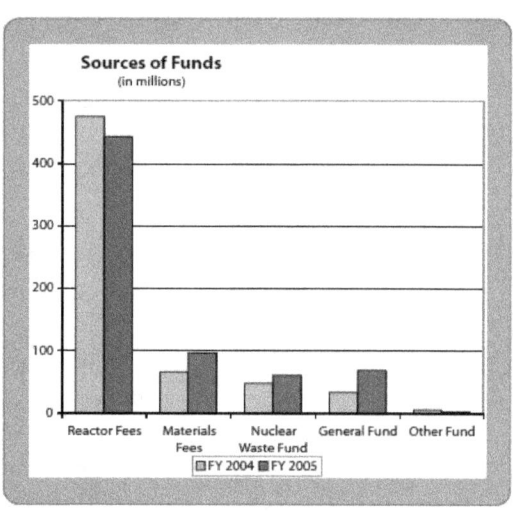

Figure 5

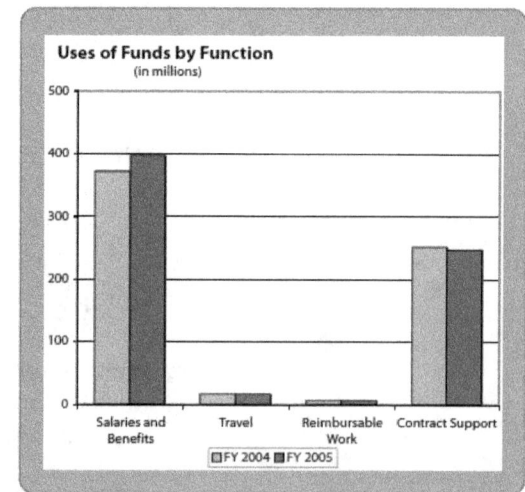

Figure 6

Financial Management Improvement Act (Improvement Act). The agency's core accounting and payroll systems, cross-serviced by the Department of the Interior's National Business Center (DOI-NBC) as part of the e-Government initiative, were also found to be in substantial noncompliance with the Improvement Act for lack of compliance with Federal financial management system requirements. The agency will continue to implement corrective actions in FY 2006 for the material weakness and substantial noncompliances.

The auditors closed two of the remaining four prior-year reportable conditions concerning user organization compensating controls for the payroll system and fee recovery from licensees. The remaining two reportable conditions concern the development of the hourly rate for license fees and accounting for internal use software. During FY 2006, the agency expects to implement corrective action for the hourly rate development for license fees and will continue to promote strengthening its internal use software practices.

Limitations of the Financial Statements

The principal statements have been prepared to report the financial position and results of operations of the NRC, pursuant to the requirements of the Chief Financial Officers Act of 1990, as amended by the Government Management and Reform Act of 1994. These statements have been prepared for the books and records of the NRC in accordance with the formats prescribed by the Office of Management and Budget. However, these statements differ from the financial reports used to monitor and control budgetary resources that are prepared from the same books and records. The principal statements should be read with the realization that they are for a sovereign entity, liabilities not covered by budgetary resources cannot be liquidated without enactment of an appropriation, and the payment of all liabilities other than for contracts can be abrogated by the sovereign entity. Other limitations are included in the footnotes to the principal statements.

The NRC's FY 2005 financial statements were audited by R. Navarro and Associates, Inc., under contract to the NRC Office of the Inspector General.

Financial Statement Highlights

The NRC's financial statements summarize the financial activity and financial position of the agency. The financial statements, footnotes, and required supplementary information, appear in Chapter 3, *Auditors' Reports and Financial Statements*. Analysis of the principal statements follows.

Analysis of the Balance Sheet

The NRC's assets were approximately $313.7 million as of September 30, 2005. This is an increase of $30.3 million from the end of FY 2004. The assets reported in NRC's Balance Sheet are summarized in the accompanying table.

The Fund Balance with Treasury represents the NRC's largest asset of $220.7 million as of September 30, 2005, an increase of $20.4 million from the FY 2004 year-end balance. This balance accounts for approximately 70 percent of total assets and represents appropriated funds, collected license fees, and other funds maintained at the U.S. Department of the Treasury to pay current liabilities. The increase in Fund Balance with the U.S. Treasury is primarily because NRC did not expend $23.0 million of the transfer from the DOE for Nuclear Waste Fund activities. Expenditures in the Nuclear Waste Fund were lower than anticipated due to delays in receipt of DOE's license application for a high-level waste repository and in procurement for the Package Performance Study. At the time the FY 2005 budget was formulated, NRC expected to receive DOE's license application in December 2004 and budgeted for increased resource needs for license review and hearings.

ASSET SUMMARY (in millions)	FY 2005	FY 2004
Fund Balance with Treasury	$220.7	$200.3
Accounts Receivable, Net	64.0	54.0
Property & Equipment, Net	27.0	26.8
Other	2.0	2.3
Total Assets	**$313.7**	**$283.4**

Accounts Receivable, Net, as of September 30, 2005, was $64.0 million and includes an offsetting allowance for doubtful accounts of $2.9 million. This is an 18 percent increase from the FY 2004 year-end Accounts Receivable, Net, balance of $54.0 million. This increase is due to an improved accounts receivable accrual methodology used in FY 2005. Accounts Receivable Due from the Public was $60.8 million, representing 19 percent of total assets. The value of Property and Equipment, Net, was $27.0 million, representing 8 percent of total assets. The majority of the balance is comprised of nuclear reactor simulators, leasehold improvements, and computer hardware and software.

The NRC's liabilities were $156.2 million as of September 30, 2005. The accompanying table shows an increase in Total Liabilities of $10.3 million from the FY 2004 year-end balance of $145.9 million. This increase is primarily in the liability to offset the increase in the unbilled accounts receivable which will be paid to the U.S. Treasury when collected. This liability increased as a result of improved estimation of accounts receivable accrual. Other Liabilities include $63.6 million for recoveries from unbilled accounts receivable, $13.0 million for accrued salaries to employees, and $33 million for accrued annual leave. Of the agency's liabilities, $43.3 million were not covered by budgetary resources, which is a slight increase over the balance as of September 30, 2004. Liabilities not covered by budgetary resources are unfunded pension expenses, accrued annual leave, and future workers' compensation. The Federal budget process does not recognize the cost of future benefits for today's employees. Instead, the Federal budget process recognizes those costs in future years when they are actually paid.

LIABILITIES SUMMARY (in millions)

	FY 2005	FY 2004
Accounts Payable	$29.0	$27.9
Federal Employee Benefits	8.4	8.1
Other Liabilities	118.8	109.9
Total Liabilities	**$156.2**	**$145.9**

The difference between total assets and total liabilities, Net Position, was $157.5 million as of September 30, 2005. This is an increase of $20.0 million from the FY 2004 year-end balance. Unexpended Appropriations is the amount of authority granted by Congress that has not been expended. The increase of Unexpended Appropriations of $20.9 million is because of the increase in Fund Balance with Treasury due to the lower expenditures from the delays in the receipt of DOE's license application for a high-level repository (see above explanation). Cumulative Results of Operations represent net results of operations since the NRC's inception. The decrease is primarily the result of a $1.3 million increase in future funding requirements related to increase in Federal Employees Compensation Act (FECA) paid by the U.S. Department of Labor.

NET POSITION SUMMARY (in millions)

	FY 2005	FY 2004
Unexpended Appropriations	$170.8	$149.9
Cumulative Results of Operations	(13.3)	(12.4)
Total Net Position	**$157.5**	**$137.5**

Analysis of the Statement of Net Cost

The Statement of Net Cost presents the net cost of NRC's two programs as identified in the NRC Annual Performance Plan. The purpose of this statement is to link program performance to the cost of programs. The NRC's net cost of operations for the year ended September 30, 2005, was $133.0 million, which is an increase of $22.6 million over the FY 2004 net cost of $110.4 million. Net costs by program are shown in the accompanying table. Gross costs increased primarily because of Federal pay raises and other nondiscretionary compensation and benefits increases.

NET COST OF OPERATIONS (in millions)		
	FY 2005	**FY 2004**
Nuclear Reactor Safety	$0.5	$(12.0)
Nuclear Materials & Waste Safety	132.5	122.4
Net Cost of Operations	**$133.0**	**$110.4**

Total exchange revenue for the year ended September 30, 2005, was $550.0 million, which is a decrease of $2.2 million from the exchange revenue of $552.2 million for the year ended September 30, 2004. Exchange revenue is derived from fees for licensing, inspections, other services, and annual fees assessed in accordance with 10 CFR Parts 170 and 171 and decreased due to the 2 percent reduction in the amount the agency was required to recover under OBRA-90.

Analysis of Statement of Changes in Net Position

The Statement of Changes in Net Position reports the change in net position during the reporting period. Net position is affected by changes in its two components—Cumulative Results of Operations and Unexpended Appropriations. The increase in Net Position of $20.0 million from FY 2004 to FY 2005 is due primarily from the net change in Unexpended Appropriations. The increase of Unexpended Appropriations of $20.9 million is because of the increase in Fund Balance with Treasury due to the lower expenditures from the delays in the receipt of DOE's license application for a high-level repository.

Analysis of the Statement of Budgetary Resources

The Statement of Budgetary Resources shows the sources of budgetary resources available and the status at the end of the period. It presents the relationship between budget authority and budget outlays, and reconciles obligations to total outlays. For FY 2005, NRC had Total Budgetary Resources available of $722.9 million, the majority of which was derived from new budget authority. This represents a 6 percent increase over FY 2004 budgetary resources available of $681.6 million. This 6 percent increase consists of an increase to fund Federal pay raises and other nondiscretionary compensation and benefits, from the Nuclear Waste Fund to initiate the review of the anticipated DOE application to construct a high-level waste repository at Yucca Mountain, and to support the Package Performance Study addressing the safety of spent nuclear fuel shipping containers in rail and highway accidents.

For FY 2005, the Status of Budgetary Resources showed the NRC incurred obligations of $665.5 million, or 92 percent of funds available. This is comparable to FY 2004 obligations of $645.3 million, or 94 percent of funds available. The decrease from 94 percent of funds available to 92 percent is because NRC did not expend $23 million of the transfer from DOE for Nuclear Waste Fund activities. Total Outlays for FY 2005 were $645.2 million, which represents a $28.6 million increase from FY 2004 total Outlays of $616.6 million.

Analysis of the Statement of Financing

The Statement of Financing is designed to provide the bridge between accrual-based (financial accounting) information in the Statement of Net Cost and obligation-based (budgetary accounting) information in the Statement of Budgetary Resources by reporting the differences and reconciling the two statements. This reconciliation ensures that the proprietary and budgetary accounts in the financial management system are in balance. The Statement of Financing takes budgetary obligations of $665.5 million and reconciles to the net cost of operations of $133.0 million by deducting non-budgetary resources, costs not requiring resources, and financing sources yet to be provided.

SYSTEMS, CONTROLS, AND LEGAL COMPLIANCE

This section provides information on NRC's compliance with the Federal Managers' Financial Integrity Act, Federal Financial Management Improvement Act, Prompt Payment Act, Debt Collection Improvement Act, Biennial Review of User Fees, Inspector General Act, and other key legal and regulatory requirements.

Federal Managers' Financial Integrity Act

The Federal Managers' Financial Integrity Act (Integrity Act) mandates that agencies establish controls that reasonably ensure that (i) obligations and costs comply with applicable law; (ii) assets are safeguarded against waste, loss, unauthorized use, or misappropriation; and (iii) revenues and expenditures are properly recorded and accounted for. This act encompasses program, operational, and administrative areas as well as accounting and financial management. It also requires the Chairman to provide an assurance statement on the adequacy of management controls and conformance of financial systems with Governmentwide standards.

INTEGRITY ACT STATEMENT

The U.S. Nuclear Regulatory Commission evaluated its management controls and financial management systems for FY 2005, as required by the Federal Managers' Financial Integrity Act. On the basis of the NRC's comprehensive management control program, I certify, with reasonable assurance, that the agency is in compliance with the provisions of this act.

Nils J. Diaz
Chairman
November 15, 2005

Management Control Review Program

Managers throughout the NRC are responsible for ensuring that effective controls are implemented in their areas of responsibilities. Each office director and regional administrator prepared an annual assurance statement that identified any control weaknesses that required the attention of the NRC's Executive Committee on Management Controls. These statements were based on various sources and included:

- Management knowledge gained from the daily operation of agency programs and reviews.

- Management reviews.

- Program evaluations.

- Audits of financial statements.

- Reviews of financial systems.

- Annual performance plans.

- Inspector General and Government Accountability Office reports.

- Reports and other information provided by the congressional committees of jurisdiction.

The NRC's Executive Committee on Management Controls is comprised of senior executives from offices of the Chief Financial Officer and the Executive Director of Operations, with the General Counsel and the Inspector General participating as advisors. The committee met and reviewed these individual assurance statements. The committee then advised the Chairman whether the NRC had any management control deficiencies serious enough to be reported as a material weakness or material noncompliance.

The NRC's ongoing management control program requires, among other things, that management control deficiencies are integrated into offices' and regions' annual operating plans. The operating plan process has provisions for periodic updates and for attention from senior managers. The management control information in these plans, combined with the individual assurance statements discussed previously, provides the framework for monitoring and improving the agency's management controls on an ongoing basis.

FY 2005 Integrity Act Results

The NRC evaluated its management control systems for the fiscal year ending September 30, 2005. This evaluation provided reasonable assurance that the agency's management controls achieved their intended objectives. As a result, management concluded that the NRC did not have any material weaknesses, as defined by the Integrity Act, in its programmatic or administrative activities. The Fee Billing System was identified as a significant management control weakness and was of sufficient importance to merit the close attention of senior management.

Federal Financial Management Improvement Act

The Improvement Act requires each agency to implement and maintain systems that comply substantially with (i) Federal financial management system requirements, (ii) applicable Federal accounting standards, and (iii) the standard general ledger at the transaction level. The act requires the Chairman to determine whether the agency's financial management systems comply with the Improvement Act and to develop remediation plans for systems that do not comply.

FY 2005 Improvement Act Results

As of September 30, 2005, the NRC evaluated its financial systems to determine if they complied with applicable Federal requirements and accounting standards required by the Improvement Act. The following eight systems were evaluated, the Federal Financial System, Federal Personnel and Payroll System, Human Resources Management System, Cost Accounting System, Advice of Allotments/Financial Plan, Capitalized Property System, Fee Billing System, and Controller Resource Database System.

The Chairman of the NRC determined that as of September 30, 2005, NRC financial management systems were in substantial compliance with the Improvement Act, except for the Fee Billing System and the Federal Financial System and Federal Personnel and Payroll System cross-serviced by the Department of the Interior's National Business Center which are in substantial noncompliance with Federal financial management system requirements. In making his determination, the Chairman considered all the information available to him, including the NRC Executive Committee on Management Control's report on the effectiveness of internal controls, Office of the Inspector General audit reports, and the results of the financial management systems reviews conducted by the agency. He also relied upon the National Business Center's annual reasonable assurance statement in which they concluded that the financial systems NRC cross-services with them, as part of the e-Government initiative, are in substantial noncompliance with Federal financial management systems requirements.

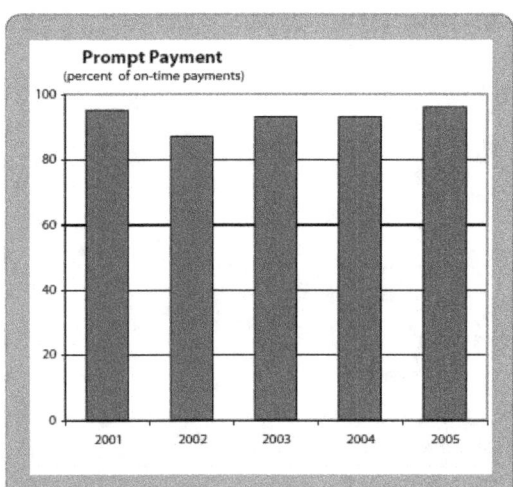

Figure 7

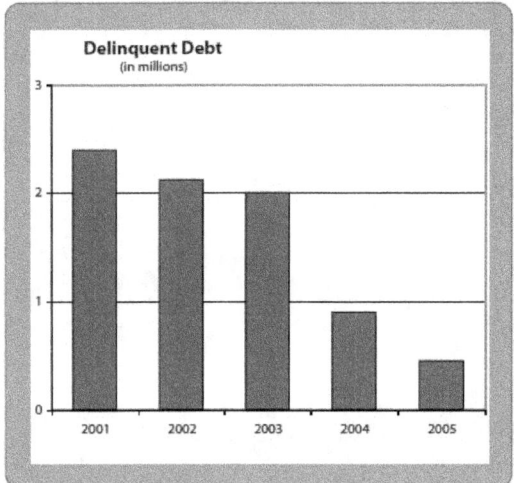

Figure 8

Prompt Payment

The Prompt Payment Act requires Federal agencies to make timely payments to vendors for supplies and services, to pay interest penalties when payments are made after the due date, and to take cash discounts when they are economically justified. During FY 2005, the NRC paid 8,629 invoices that were subject to the Prompt Payment Act. The NRC increased its percentage of on-time payments subject to the Prompt Payment Act from 94 percent in FY 2004 to 96 percent in FY 2005 (see Figure 7). The amount of interest penalties incurred during FY 2005 was $8,850, which is an increase from FY 2004's amount of $2,917.

Debt Collection

The Debt Collection Improvement Act is intended to enhance the ability of the Federal Government to service and collect debts. The agency's goal is to maintain the delinquent debt owed to the NRC, at year-end, to less than one percent of its annual billings. The NRC continues to meet this goal and has kept delinquent debt to less than one percent for the past 10 years. Delinquent debt at the end of FY 2005 was $0.45 million (see Figure 8). This is a decrease of $0.4 million over FY 2004 and a decrease in the number of outstanding receivables from 154 to 96. The NRC continues to pursue the collection of delinquent debt and continues to timely refer all eligible delinquent debt over 180 days to the U.S. Treasury for collection.

Biennial Review of User Fees

The Chief Financial Officers Act requires agencies to conduct a biennial review of fees, royalties, rents, and other charges imposed by agencies, and make revisions to cover program and administrative costs incurred. Each year, the NRC revises the hourly rates for license and inspection fees and adjusts the annual fees to meet the fee collection requirements of the Omnibus Budget Reconciliation Act of 1990, as amended. The most recent changes to the license, inspection, and annual fees are described in the *Federal Register* (70 FR 30526, May 26, 2005).

During FY 2005, the NRC reviewed its fees for criminal history licensees, public use of the auditorium, small materials licenses, international program materials licenses, and import/export licenses subject to the biennial review requirement. Reviews of other types of fees concluded that revisions were not warranted at this time.

Treasury Performance Measure Summary

Treasury has four key performance indicators for measuring how agencies complied with reporting requirements for the Governmentwide Financial Reporting System (GFRS), the Federal Agencies Centralized Trial Balance System (FACTS I), and intragovernmental activity. Overall for FY 2004, the NRC complied with the four performance indicators for timely reporting, reconciliation of unexplained differences for intragovernment activity, reliability and completeness of intragovernment reporting, and consistency and reasonableness. Treasury has not issued its FY 2005 Performance Measure Summary; however, based on our self-evaluation, NRC also met the requirements for this fiscal year.

Inspector General Act

The agency has established and continues to maintain an excellent record in resolving and implementing open audit recommendations presented in Office of the Inspector General reports. Section 5(b) of the Inspector General Act requires agencies to report on final actions taken on audit recommendations. This information as well as data concerning disallowed costs determined through contract audits conducted by the Defense Contract Audit Agency can be found in Appendix C.

Improper Payments

Improper payments continue to be at low risk for the agency. The NRC continues to evaluate its internal controls to guard against improper payments and monitors and reports on improper payments within its programs. At the present time, NRC's inventory of functional payment areas consists of commercial vendor, interagency, and travel payments. The DOI-NBC's Federal Personnel Payroll System became the NRC system of record for payroll disbursements effective November 2, 2003. The DOI-NBC is responsible for monitoring and reporting on any improper payroll-related payments. The NRC will continue to perform annual risk assessments for each of these areas. Based on the FY 2005 risk assessments, the number of and amount of improper payments fall below external reporting requirements established by Office of Management and Budget guidance on what is considered to be a significant risk.

"Getting to Green" for the five management initiatives . . .

MEASURING AND REPORTING PERFORMANCE

This chapter presents information on performance measurement at the NRC during FY 2005. The agency's performance measures are reported by each of the five goals contained in the NRC's FY 2004–FY 2009 Strategic Plan. The goals in the Strategic Plan focus on safety, security, openness, effectiveness, and management excellence. Strategic outcomes under each goal define success in meeting each Strategic Plan goal. Further, the performance measures associated with each goal indicate how effectively the NRC is achieving its goals and establish the basis for performance management.

This chapter also describes the achievements and challenges faced by each activity under the agency's major programs of Nuclear Reactor Safety and Nuclear Materials and Waste Safety. The activities under the agency's major programs consist of Nuclear Reactor Licensing, Nuclear Reactor Inspection, Fuel Facilities Licensing and Inspection, Nuclear Materials Users Licensing and Inspection, High-Level Waste Repository, Decommissioning and Low-Level Waste, and Spent Fuel Storage and Transportation Licensing and Inspection.

Following the program-specific discussions, the NRC's progress in "Getting to the Green" for the five management initiatives identified in the President's Management Agenda is described. Lastly, information on data sources, data quality, and the completeness and reliability of the performance data are presented. The discussion focuses primarily on the methods the NRC used to collect and analyze data, ensure data security, and improve the agency's performance measures and data during the current reporting period. Endnotes are referenced throughout the document and located at the end of the report.

The performance measures reported on the following pages reflect measures contained in both the FY 2005 and FY 2006 Performance Budgets. A number of the performance measures contained in the FY 2005 Performance Budget are discontinued after FY 2005 to reflect the evolutionary change in the goals contained in the agency's FY 2004–FY 2009 Strategic Plan. Although those performance measures will not be reported in future *Performance and Accountability Reports*, the agency activities that they represent will continue to be reflected in the report. For example, although there will no longer be a performance measure reflecting the number of medical events that occur each year, this function will continue to be supported by the Commission and are reflected under the Safety and Security goals of the Strategic Plan.

GOALS AND PERFORMANCE MEASURES

Goal 1 — Safety: Ensure Protection of Public Health and Safety and the Environment

Strategic Outcomes

The NRC has five strategic outcomes associated with the Safety goal that determine whether the agency has achieved its goal to ensure protection of public health and safety and the environment.

1.1   No nuclear reactor accidents.[1]

1.2   No inadvertent criticality events.

1.3   No acute radiation exposures resulting in fatalities.

1.4   No releases of radioactive materials that result in significant radiation exposures.

1.5   No releases of radioactive materials that cause significant adverse environmental impacts.

Results: In FY 2005, the NRC achieved all of its Safety goal strategic outcomes.

Performance Measures

In the following table are the FY 2005 Safety goal performance measures and targets stated in the FY 2005 and FY 2006 Performance Budgets. The NRC met all of the FY 2005 Safety goal performance measure targets.

Goal 1: FY 2005 Safety Performance Measures in the FY 2005 and FY 2006 Performance Budgets				
Measure	2002	2003	2004	2005
1. Number of new conditions evaluated as red by the reactor oversight process[2] is ≤ 3.	New metric in FY 2005			0
2. Number of significant accident sequence precursors of a nuclear reactor accident is 0.[3]	1	0	0	0
3. Number of operating reactors whose integrated performance entered the Manual Chapter 0350 process, the multiple/repetitive degraded cornerstone column or the unacceptable performance column of the Reactor Oversight Program Action Matrix with no performance exceeding Abnormal Occurrence Criterion I.D.4 is ≤ 4.[4]	New metric in FY 2005			0
4. Number of significant adverse trends in industry safety performance with no trend exceeding the Abnormal Occurrence Criterion I.D.4 is ≤ 1.[5]	0	0	0	0
5. Number of events with radiation exposures to the public and occupational workers that exceed Abnormal Occurrence Criterion I.A.				
Reactors 0	0	0	0	0
Materials ≤ 6	0	0	0	1
Waste 0	0	0	0	0
6. Number of radiological releases to the environment that exceed applicable regulatory limits.[6]				
Reactors ≤ 3[7]	0	0	0	0
Materials ≤ 5	4	0	1	0
Waste 0	0	0	0	0
7. No more than 300 losses of control[8] of licensed material per year.[9]	272	219	201	211
8. No more than 30 events per year [10] resulting in radiation overexposures [11] from radioactive material that exceed applicable regulatory limits.	23	16	4	16
9. No more than 45 medical events per year.[12]	33	39	40	32
10. No non-radiological events that occur during NRC-regulated operations that cause impacts on the environment that cannot be mitigated within applicable regulatory limits, using reasonably available methods.[13]	0	0	0	0

Analysis of Results

1. **Reactor Oversight Process Red Conditions:** The NRC monitors nuclear power plant performance in three broad areas: reactor safety, radiation safety, and security. The NRC has established the following categories: green, white, yellow, and red, with red being the category of highest significance. Red findings indicate a significant reduction in the safety of a nuclear power plant. There were no red performance indicators or findings in FY 2005.

2. **Significant Precursors:** The second measure tracks significant precursor events, defined as those events that have a probability of 1 in 1,000 or greater of leading to substantial damage to the reactor fuel. In FY 2005, the performance measure threshold was reduced from no more than one significant precursor event per year to no significant precursor events during the year. There were no significant precursor events in FY 2005.

3. **Reactor Performance:** The conditions in this measure indicate that significant issues are found in a plant during inspections conducted under the Reactor Oversight Program. All of the conditions in this measure indicate NRC action will be undertaken based on the findings. There were no reactors that met the conditions in this measure in FY 2005.

4. **Adverse Safety Trends:** This measure tracks the trends of several key indicators of industry safety performance. The indicators provide insights into major areas of reactor performance, including reactor safety, radiation safety, and physical protection. These trends represent industry averages, rather than individual plant performance. Statistical analysis techniques are applied to each indicator to determine its long-term trend. To date, there have been no statistically significant adverse trends in any of the indicators. The data are current as of June 1, 2004.

5. **Radiation Exposures:** Number of events with radiation exposures to the public and occupational workers that exceed Abnormal Occurrence Criterion I.A. There was one radiation exposure exceeding Abnormal Occurrence I.A.

6. **Releases to the Environment:** This measure is an indicator of the effectiveness of the NRC's nuclear materials environmental programs. The industry had no releases to the environment that exceeded regulatory limits in FY 2005.

7. **Losses of Control:** This measure tracks reportable events of materials entering the public domain in an uncontrolled manner. The industry experienced a total of 211 losses of control of licensed material in FY 2005 through September 30, 2005. Many of the events counted toward this measure do not, by themselves, pose a risk to public health and safety. For example, most of the losses of control of licensed material involve shielded materials, which are unlikely to result in overexposures to individuals or releases to the environment with most eventually recovered. However, these losses are tracked because they may indicate weaknesses in licensees' programs. Very few of the events tracked involve high enough quantities of radioactive material to pose a security concern. This measure has been superseded in FY 2005 by a new security measure that tracks the number of events involving unrecovered risk-significant material. This measure is discontinued after FY 2005.

8. **Radiation Overexposure:** The industry experienced 16 events in FY 2005 that resulted in radiation overexposures from radioactive material that exceeded applicable regulatory limits through September 30, 2005. For fuel cycle facilities, this measure extends to other hazardous materials that are used with, or produced from, licensed material, consistent with 10 CFR Part 70. Reportable chemical exposures are those that exceed license commitments. They also include chemical exposures involving uranium recovery activities under the Uranium Mill Tailings Radiation Control Act. This measure is discontinued after FY 2005.

9. **Medical Events:** The industry experienced 32 medical events in FY 2005 through September 30, 2005. Since data collection began under the Government Performance and Results Act, the peak year was FY 1998, when 42 events occurred. This measure pertains to medical events reported under 10 CFR Part 35, "Medical Use of Byproduct Material." The NRC's Medical Use Program includes those who use byproduct material in medical diagnosis and therapy. This measure is discontinued after FY 2005.

10. **Nonradiological Events:** The industry did not experience any nonradiological events during NRC-regulated operations that had an impact on the environment during FY 2005. This measure involves only chemical releases from the uranium mining and milling facilities that are regulated by the NRC. Examples of events that might be counted include chemical releases resulting from excursions at onsite leach facilities or releases from mill tailings piles that could contaminate groundwater. This measure is discontinued after FY 2005.

Goal 2 — Security: Ensure the Secure Use and Management of Radioactive Materials

Strategic Outcome

The NRC has one strategic outcome associated with this goal that determines whether the agency has achieved its goal to ensure the secure use and management of radioactive materials:

2.1   No instances where licensed radioactive materials are used domestically in a manner hostile to the security of the United States.

Results: In FY 2005, the NRC achieved the Security goal strategic outcome.

Performance Measures

In the following table are the FY 2005 Security goal performance measures and targets stated in the FY 2005 and FY 2006 Performance Budgets. The NRC met all of the FY 2005 Security goal performance measure targets.

GOAL 2: FY 2005 Security Performance Measures in the FY 2005 and FY 2006 Performance Budgets				
Measure	**2002**	**2003**	**2004**	**2005**
1. Unrecovered losses or thefts of risk-significant radioactive sources is zero.	0	0	0	0
2. Number of security events and incidents that exceed the Abnormal Occurrence Criteria I.C. 2-4. is ≤ 4.	New metric in FY 2005			0
3. Number of significant unauthorized disclosures of classified and/or safeguards information is zero.[14]	0	0	0	0

Analysis of Results

1.   **Unrecovered Losses or Thefts:** This measure covers any loss or theft of radioactive nuclear material that the NRC has determined to be risk significant. There were no losses or thefts of risk significant radioactive material in FY 2005.

2.   **Security Events:** This measure covers substantiated cases of actual or attempts of theft or diversion of licensed nuclear material or sabotage of a nuclear facility, any substantiated loss of special nuclear material, or any substantiated inventory discrepancy judged to be significant relative to normally expected performance

and that is judged to be caused by theft or diversion or by substantial breakdown of the accountability system. Substantiated means a situation where an indication of loss, theft or unlawful diversion such as an allegation of diversion, report of lost or stolen material, statistical processing difference, or other indication of loss of material control or accountability cannot be refuted following an investigation; and requires further action on the part of the agency or other proper authorities. There were no events that met the conditions for this measure in FY 2005.

3.   **Significant Disclosures:** Significant unauthorized disclosures of classified and/or safeguard information that cause damage to national security or public safety. There were no documented disclosures during FY 2005.

Goal 3 — Openness: Ensure Openness in Our Regulatory Process

Strategic Outcome

The NRC has one Strategic Outcome associated with this goal that is used to determine whether the agency has achieved its goal to ensure openness in our regulatory processes.

3.1   Stakeholders are informed and involved in NRC processes as appropriate.

Results: The NRC met the Openness strategic outcome target in FY 2005.

Performance Measures

In the following table are the FY 2005 Openness goal performance measures and targets stated in the FY 2005 and FY 2006 Performance Budgets. The NRC met all of the FY 2005 Openness goal performance measure targets except the first performance measure.

GOAL 3: FY 2005 Openness Performance Measure in the FY 2005 and FY 2006 Performance Budgets				
Measure	**2002**	**2003**	**2004**	**2005**
1. Percentage of selected openness output measures that achieve performance targets is ≥70%.	New measure in FY 2005			50%
A. Respond to Freedom of Information Requests < 20 days.				12
B. Issue 90 percent of Directors Decisions under 2.206 within 120 days.				100%
C. Make 85 percent of Final Significant Determination Process Determinations within 90 days for all potentially greater than Green Findings.				68%
D. At least 90 percent of Category 2 and 3 meetings on regulatory issues for which public notices are issued 10 days in advance of the meeting.				89%
2. Complete milestones related to collecting, analyzing, and trending information for measuring public confidence.	Met	Met	Met	Met
3. Complete all public outreaches.				
Reactors	Met	Met	Met	Met
Materials	Met	Met	Met	Met
Waste	Met	Met	Met	Met
4. Issue Director's Decisions for petitions filed to modify, suspend, or revoke a license under 10 CFR 2.206 within an average of 120 days.	Number of Days:			
Reactors	126	115	88	NA*
Materials	NA*	NA	48	119
Waste	167	115	NA	NA
* NA – None filed				

Analysis of Results

1.   **Openness Output Measures:** This measure is based on the following output measures:

    (A)   Respond to Freedom of Information Act Requests in 20 days or less: This measure tracks the NRC's responsiveness to an important type of public request for information, Freedom of Information Act Requests. In FY 2005, the median number of days for responding to FOIA requests was 12.

    (B)   Issue 90 percent of Director's Decisions under 2.206 within 120 days: This measure tracks the NRC's responsiveness to a special type of public request for information, Director's Decisions. 10 CFR 2.206 gives individuals an opportunity to file a petition to institute a proceeding to modify, suspend, or revoke a license or for any other action that may be proper. All of the Directors Decisions were issued within 120 days.

    (C)   Make 85 percent of Final Significance Determination Process Determinations within 90 days for all potentially greater than green findings: This measure tracks the timeliness of Significance Determination Process determinations. The Agency did not meet this output target. Only 50 percent of final Significance Determination Process determinations were made within 90 days of all potentially Greater than Green findings. This was due to the closure of numerous late Significance Determination Process issues, mainly associated with fire-related inspection findings. However, with increased management focus and several programmatic changes in reactor inspection activities, it is anticipated that significant improvements will be made in FY 2006.

    (D)   At least 90 percent of Category 2 and 3 meetings on regulatory issues for which public notices are issued at least 10 days in advance of the meeting: This measure tracks the timeliness with which the NRC notifies the public of meetings. In FY 2005, the NRC issued 89 percent of Category 2 and 3 meeting notices at least 10 days in advance of the meeting date. OIS provided the offices with quarterly statistics showing their performance under this measure. However, FY 2005 is the first year that offices measured and reported their performance against this goal in their operating plans. During the first and second quarters of FY 2005, the offices' failure to meet the performance measure was attributable to the following: (1) notices were not submitted in time, and/or (2) notices were not declared and replicated in ADAMS before submission. As a result, OIS worked closely with the offices to assist them in reaching the performance measure in the third and fourth quarters of FY 2005.

2. **Public Confidence:** This measure contains a series of milestones to be used to measure public confidence. A public meeting feedback form was used to collect comments on the effectiveness of the public meetings in building confidence in the NRC and institute changes where warranted. A database was developed to more accurately and efficiently track the responses to the meeting forms. The milestones for collecting, analyzing, and trending this information were met. This measure is discontinued after FY 2005.

3. **Public Outreaches:** Outreach meetings were held during the year to provide the public with opportunities for meaningful participation in NRC activities. All of the scheduled meetings were held. This measure is discontinued after FY 2005.

4. **Directors Decisions:** 10 CFR 2.206 gives individuals an opportunity to file a petition to institute a proceeding to modify, suspend, or revoke a license or for any other action that may be proper. All of the petitions that were filed were addressed within an average of 120 days in FY 2005. One petition, the Sequoyah Fuels petition, is still pending from FY 2004. It relates to a licensing review, and the Director's Decision cannot be developed until the review is completed. This measure is discontinued after FY 2005.

Goal 4 — Effectiveness: Ensure that NRC Actions are Effective, Efficient, Realistic, and Timely

Strategic Outcome

The NRC has one strategic outcome associated with this goal that determines whether the agency has achieved its goal to ensure that NRC actions are effective, efficient, realistic, and timely:

> 4.1   No significant licensing or regulatory impediments to the safe and beneficial uses of radioactive materials.

Results: NRC failed to achieve the Effectiveness strategic outcome because the PART performance measure associated with this goal was not met (see below).

Performance Measures

In the following table are the FY 2005 Effectiveness goal performance measure and targets stated in the FY 2005 and FY 2006 Performance Budgets. The NRC met all of the FY 2005 Effectiveness goal performance measure targets except for the goals to receive a minimum score of 85 from the Office of Management and Budget on programs assessed using the Program Assessment Rating Tool (PART)

Goal 4: FY 2005 Effectiveness Performance Measures in the FY 2005 and FY 2006 Performance Budgets				
Measure	**2002**	**2003**	**2004**	**2005**
1. Programs assessed during the fiscal year using the Program Assessment Rating Tool (PART) receive a minimum score of 85 from OMB:	New measure in FY 2005			
Reactor Licensing				74%
Spent Fuel Storage and Transportation Licensing and Inspection				89%
2. Complete specific milestones in the Risk-Informed Regulation Implementation Plan.				
Reactors	Met	Met	Met	Not Met
Materials	Met	Met	Met	Met
Waste	Met	Met	Met	Met
3. Complete at least two key process improvements per year in selected program and support areas that increase effectiveness, efficiency, and realism.				
Reactors	2	2	2	2
Materials	2	3	2	2
Waste	2	2	3	2
4. Complete those major milestones scheduled in accordance with the Commission-approved schedules in order to support completion of license renewal applications within 30 months from receipt of application to a Commission decision if a hearing is held (within 22 months without a hearing).				
Reactors	Met	Met	Met	Met
5. Complete those specific milestones to reduce unnecessary regulatory burden.				
Reactors	Met	Met	Met	Met
Materials	Met	Met	Met	Met
Waste	Met	Met	Met	Met

Analysis of Results

1. **PART Results:** The NRC did not meet the performance measure target. The Reactor Licensing activity received a score of 74, or "moderately effective" on the Program Assessment Rating Tool evaluation. The Office of Management and Budget has recommended that the NRC eliminate this measure from the Performance Budget because detailed Program Assessment Rating Tool findings, rather than the program's rating by itself, inform budget recommendations. The NRC will explore whether to delete this performance measure beginning in the FY 2007 Performance Budget.

2. **Risk-Informed Regulation Implementation Plan:** All of the milestones in the Risk-Informed Regulation Implementation Plan for materials and waste were met. The NRC completed all of the nuclear reactor safety milestones in the Risk-Informed Regulation Implementation Plan (RIRIP) on schedule except for one. The milestone "Issue final regulatory guide for the risk-informed performance-based fire protection rule" was rescheduled to November 2005 due to a delay in the receipt of NEI information and resolution of ACRS comments.

    The milestones include publication of the mitigating systems performance index pilot verification report, reevaluation of station blackout risk using upgraded risk models, completion of an expert elicitation process supporting a proposed rule change for loss of coolant accidents, and publishing of a report on good practices for human reliability analysis. This measure is discontinued after FY 2005.

3. **Process Improvements:** There were 2 process improvements each for the NRC's reactor, material, and waste activities. The process improvements for reactor activities were for allegation and investigation processes. Allegation activities were improved by utilizing voluntary alternative dispute resolution to assist in the resolution of discrimination complaints for engaging in protected activity. The second process improvement was to initiate an integrated case system that tracks allegations, enforcement, and investigation information. Process improvements for materials and waste safety activities included the development of a decision-making framework for security assessments to evaluate and prioritize the need for additional security measures. A notable process improvement was the application of more realistic dose modeling scenarios to accelerate license termination reviews, resulting in the completion of the Yankee Rowe License Termination Plan review in approximately half the average time taken for such a review. The other process improvements were to develop a database to enable the agency to track information for waste disposals

authorized in accordance with 10 CFR 20.2002 and the initiation of an effort focused on continuous improvement in the by product materials program. This measure is discontinued after FY 2005.

4. **License Renewal Applications:** This measure is to ensure that the NRC handles license renewal reviews in an expeditious manner. As of May 31, 2005, the NRC completed license renewal reviews for 10 units in FY 2004 and for 6 units in FY 2005. All 16 renewed licenses were issued within the target time frame of 30 months with a hearing, or 22 months without a hearing. This measure is discontinued after FY 2005.

5. **Unnecessary Regulatory Burden:** Regulatory proceedings and activities are evaluated for timely actions and feedback from licensees to ensure that the NRC is reducing any unnecessary burden of licensees. All of the milestones were met in FY 2005. For example, the NRC completed a final rule that amended the requirements for training and experience in 10 CFR Part 35, "Medical Use of Byproduct Material." The rule amended the regulations governing the requirements for recognition of certain specialty boards whose certification may be used to demonstrate the adequacy of the training and experience of individuals to serve as authorized medical physicists, authorized nuclear pharmacists, radiation safety officers, or authorized users of byproduct material (physicians). The rule reduces regulatory burden by making requirements more flexible. This measure is discontinued after FY 2005.

Goal 5 — Management: Ensure Excellence in Agency Management to Carry Out the NRC's Strategic Objective

Strategic Outcomes

The NRC has two strategic outcomes associated with this goal that determines whether the agency has achieved its goal to ensure the excellence in Agency Management.

5.1 Continuous improvement in NRC's leadership and management effectiveness in delivering the mission.

5.2 A diverse, skilled workforce and an infrastructure that fully supports the agency's mission and goals.

Performance Measures

In the following table are the FY 2005 Management performance measure and targets stated in the FY 2006 Performance Budget. The NRC failed to achieve the first Management strategic outcome because the performance measure target associated with this goal was not met.

Goal 5: FY 2005 Management Performance Measures in the FY 2006 Performance Budget				
Measure	**2002**	**2003**	**2004**	**2005**
1. Percentage of selected NRC management programs that deliver intended outcomes is ≥70%.	New Measure in FY 2005			60%
A. Infrastructure Management				100%
B. Financial Management and Budget & Performance Integration				67%
C. Expanded Electronic Government				50%
D. Recruitment and Staffing				80%
E. Internal Communications				100%

Analysis of Results

1. **Infrastructure Management:** Infrastructure management activities maintain a healthy, safe, secure, and accessible work environment as well as providing equipment, facilities, and administrative services that employees need. 100 percent of the infrastructure management activities achieved their performance targets.

2. **Financial Management & Budget and Performance Integration:** Financial management activities provide accurate, timely, and useful financial information to managers for decisionmaking and ensure that the NRC's financial assets are adequately protected consistent with risk. Budget and Performance Integration activities improve the linkage of individual and organizational performance standards to the NRC's Performance Budget and uses and improves the Planning, Budgeting, and Performance Management process to ensure better integration of performance results into NRC planning and budgeting. The NRC did not meet its target for not receiving any material weaknesses on the

audit of the agency's financial statement nor for having all its financial systems meet Governmentwide requirements. The material weakness was associated with inadequate acceptance testing, intensive manual processes, and the lack of comprehensive quality assurance procedures for the Fee Billing System.

The agency has implemented the following improvements to the fee billing process during FY 2005: (1) revised the quality assurance procedures for the Part 170 billing process to include a global reconciliation of each quarterly fee billing cycle; (2) modified the Fee Billing System to improve functionality of the system's interface; (3) expanded the acceptance testing for Fee Billing System software modifications; (4) performed independent verification and validation of the acceptance testing for the Fee Billing System Software Modifications; (5) separated the performance of the fee billing process and global reconciliation functions; (6) hired a staff accountant to further improve internal controls over the billing process; and (7) developed a statistical sampling plan to test that internal controls are functioning as intended.

The NRC's Fee Billing System and the payroll and core accounting systems cross-serviced by the Department of Interior's National Business Center are in substantial noncompliance with Federal financial management system requirements. During FY 2005, the NRC developed a remediation plan for the Fee Billing System. The plan describes our approach for overcoming the deficiencies that resulted in the substantial noncompliance. The plan includes a feasibility assessment of bringing the Fee Billing System into compliance with the Improvement Act and includes the milestones and schedule to replace the Fee Billing System with a compliant system. The Fee Systems Replacement Project is scheduled to be completed by December 2007.

In support of the Federal e-Government effort, the NRC's payroll (Federal Personnel and Payroll System) and core accounting (Federal Financial System) financial systems are cross-serviced by the Department of Interior's National Business Center. The Department of Interior Inspector General recently conducted a multi-phased penetration test of the strengths and weaknesses of the Department of Interior's and the National Business Center's networks and systems architecture. Based on the Department of Interior's Inspector General findings, the National Business Center concluded that the National Business Center's information technology systems have serious weaknesses in complying with some provisions of Appendix III of the Office of Management and Budget Circular A-130, *Management of Federal Information Resources.* As a result, the National Business Center concluded that they do not substantially comply with the improvement Act requirements. The National Business Center will need to develop a remediation plan to bring their systems into compliance with the Improvement Act. We plan to monitor the Department of Interior's progress in addressing the security issues.

3. **Expanded Electronic Government:** The NRC's overall target for Expanded Electronic Government in FY 2005 was to achieve a "yellow" rating on OMB's e-Gov scorecard. The NRC's Capital Asset Plans and Business Cases (Exhibit 300) submissions were rated very highly, receiving scores of four or five out of a maximum possible of five. Performance criteria for major systems as reflected in achieving cost and schedule goals were achieved. The NRC has institutionalized procedures to avoid duplication of its IT investments using e-Gov Presidential Priority initiatives and the Lines of Business initiatives. The NRC has also put in place procedures to report progress of e-Gov initiatives as required for "high risk" IT investments. Additionally, NRC has in place an effective agencywide IT Security Plan of Action and Milestone remediation process, verified by its Inspector General. The NRC did not achieve an OMB rating of "3" on its Enterprise Architecture, which is required to achieve a "yellow" rating on the e-Gov Scorecard. Resource considerations have required a reassessment of the rate at which the agency will be able to address issues regarding Enterprise Architecture. Aside from NRC's OMB scorecard goal, the other activity that did not meet its target was the percentage of Federal Information Security Management Act compliance across all NRC major applications and general support systems. While information at the NRC is secure, changing requirements under the Act resulted in a compliance rate of 70 percent rather than the NRC's target of 90 percent.

4. **Recruitment and Staffing:** Recruitment and staffing seeks to, among other tasks, use innovative recruitment, development and retention strategies to achieve a high quality, diverse workforce with the skills needed to achieve the agency's mission. 80 percent of recruitment and staffing activities met their targets.

5. **Internal Communications:** The agency's internal communications activities are intended to foster and support a culture of openness and innovation. All of the internal communications activities achieved their performance targets.

NUCLEAR REACTOR SAFETY

Overview

The Nuclear Reactor Safety program ensures that civilian nuclear power reactors and test and research reactors, are operated in a manner that adequately protects the public health and safety and the environment while safeguarding special nuclear materials used in reactors. The NRC regulates 104 nuclear power reactors and 35 test and research reactors that are currently licensed to operate. Nuclear power plants generate approximately 20 percent of the Nation's electricity, and test and research reactors are used to conduct research and development. Almost every field of science (including physics, chemistry, medicine, and biology) uses these reactors.

The Commission's health and safety regulations provide reasonable assurance of adequate protection of public health and safety. These regulations are based on defense-in-depth principles and conservative practices that provide an adequate margin of safety. The collective efforts of the NRC and the nuclear industry are needed in order to maintain safety. The NRC establishes rules, safety standards, and requirements for licensees; conducts thorough in-depth technical reviews of both reactor designs and the safety envelope of licensed operations; oversees safe plant operations; and responds to licensees and other stakeholders. The NRC's licensees are responsible for designing, constructing, and operating nuclear reactors safely.

Ensuring the Safe Operation of Nuclear Reactors

The NRC ensures the safety of nuclear reactors by establishing the related safety standards and requirements and conducting in-depth technical reviews in the course of licensing nuclear power plants and their operators. The NRC also oversees plant operating performance and evaluates security and emergency response activities, establishes clear health and safety regulations, conducts research to resolve safety issues, and provides technical support for developing regulations. Nuclear plant licensees are required to follow the NRC's regulations specifying how plants are to be designed, constructed, and operated.

The NRC provides independent oversight of the plants through the Reactor Oversight Process to verify that NRC licensees are operating their plants safely and in accordance with the NRC's rules and regulations. If violations are found, the NRC may take enforcement actions. Security and emergency response actions ensure that licensees take adequate measures to respond to malevolent actions against reactors and that public safety measures are in place in the event that an incident occurs. Research actions analyze data from operations and independently undertakes studies that provide the basis for maintaining the safety of nuclear power plants. The following sections describe these safety activities in greater detail.

Nuclear Reactor Licensing Activity

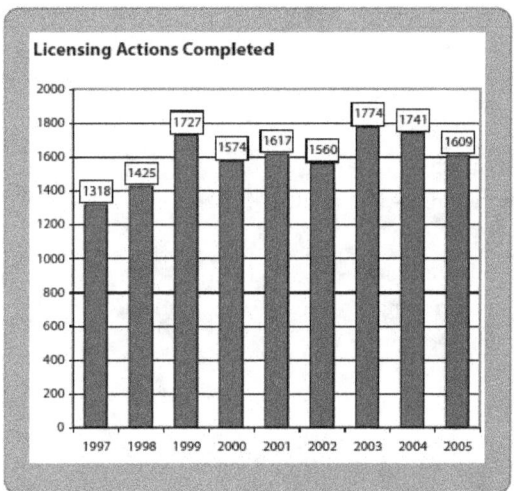

Figure 1

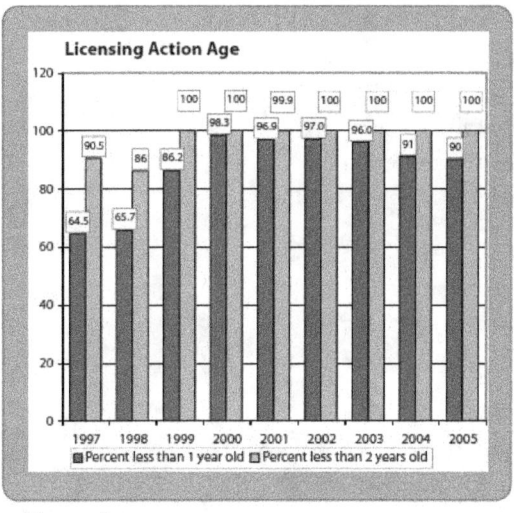

Figure 2

The Reactor Licensing activities establish requirements for licensees that set expectations for the commercial use of radioactive material within the legal framework of the NRC's safety and environmental regulations. This includes assurances that facilities are adequately designed, properly constructed, and correctly maintained, and that trained and qualified operating and technical support personnel can prevent or cope with accidents and other threats to public health and safety. The NRC also reviews license applications and changes to existing licenses, examines and licenses reactor operators, reviews reactor events for safety significance, and improves safety regulations and guidance.

The NRC met three of its five output measures for reactor licensing during FY 2005. The goals successfully achieved were completing a minimum of 1500 reactor licensing actions (see Figure 1) (1609 licensing actions were completed as of September 2005), completing a minimum of 500 other licensing tasks (715 other licensing tasks were completed as of September 2005) and maintaining a working inventory of 1200 or less licensing actions. The missed output measures were completing 100 percent of licensing actions within two years (see Figure 2) and completing extended power uprates within either 12 months or by the licensee's need date (whichever is greater). Both output measures were missed as a result of a single licensing action, the Vermont Yankee extended power uprate. In keeping with the NRC's safety goal, the Vermont Yankee extended power uprate could not be completed within the established output metrics because of issues with the licensee's steam dryer analyses and thermal hydraulic analyses. The targets for licensing action inventory and completion of licensing actions within one year by the end of the fiscal year were revised for FY 2005 to reflect redirection of resources within FY 2004 to higher priority security work including review of security plans, safeguards contingency plans, and training and qualification plans.

As of September 2005, 50 initial operator licensing examinations were given in FY 2005, thereby satisfying facility demand for new operators. Two exams were rescheduled or delayed into 2006 based upon facility requests; however, two additional retake examinations were administered to satisfy facility requests. Four Generic Fundamentals Examinations were administered in FY 2005, achieving the target of three exams.

Power Uprates

Since the 1970s, licensees have been applying for and implementing power uprates as a means of increasing the power output of their plants. The NRC's comprehensive reviews of an application are focused on the potential impacts that the proposed power uprate might have on the existing licensing-basis analyses that demonstrate overall plant safety. The review of a power uprate application assures that the impacts of increasing a plant's power output are fully addressed and that plant operation at the increased power level is safe. Power uprates increased the Nation's electrical generating capacity by approximately 234 MWe in FY 2005.

The NRC has set timeliness standards for these reviews in order to ensure a stable and predictable regulatory environment for the safety and environmental review of these licensing actions.

New Reactor Licensing

The NRC continues to focus on new reactor licensing activities to ensure that the Commission's safety requirements and expectations will be met for future reactors and a stable and predictable framework will exist for potential future license applicants. These activities are in response to the nuclear industry's continued interest in new reactors and the Department of Energy's ongoing efforts to cost-share new reactor licensing projects.

The NRC issued a final safety evaluation report and final design approval for the Westinghouse AP1000 advanced reactor design in September 2004. The rulemaking is scheduled to be completed in December 2005, but major elements of the design have been deferred by the vendor.

The NRC is actively engaged in pre-application reviews of General Electric's Economic Simplified Boiling-Water Reactor and Framatome ANP's EPR designs. The design certification application for the Economic Simplified Boiling-Water Reactor was received

August 24, 2005, however, the application has not yet been docketed. The NRC will determine the schedule for the review once the application has been accepted and docketed. The EPR design certification is expected at the end of 2007.

The NRC conducted research activities to support the pre-application review of the ACR-700 and Economic Simplified Boiling-Water Reactor new reactor designs. For example, the NRC issued an ACR-700 Phenomena Identification and Ranking Techniques report which identified and ranked by importance issues and phenomena and addressed the adequacy of the database for licensing the ACR-700 reactor. Other research activities supported completion of the ACR-700 Pre-application Safety and Assessment Report and the AP1000 Design Certification Safety Evaluation Report.

In September and October 2003, the NRC received early site permit applications for the Clinton, North Anna, and Grand Gulf sites. The NRC continued reviewing the three applications in FY 2005, and expects to complete its review and issue a decision in FY 2007.

The NRC is currently reviewing industry guidance for preparing a Combined Operating License application. The guidance was provided by the Nuclear Energy Institute in its December 2004 submittal, NEI-04-01, "Industry Guidelines for Combined License Applicants Under 10 CFR Part 52."

In June 2005, the NRC issued Inspection Manual Chapter 2502, "Construction Inspection Program: Pre-Combined License Phase," describing the inspections the NRC will conduct during the review of a Combined Operating License application. The NRC also issued nine implementing procedures to support the inspection of quality assurance and the environmental impact of site preparation work. The manual chapter and supporting inspection procedures complete the inspection infrastructure to support the review of a Combined Operating License application.

The NRC continues to develop the regulatory infrastructure needed to inspect new reactor and site license applications and do effective and efficient licensing reviews of those applications. Toward that end, the NRC is currently considering stakeholder comments received in response to proposed revisions to the regulation governing early site permits, design certifications, and combined licenses. The NRC is continuing its interaction with industry representatives on generic issues associated with the receipt of a combined license application. These actions are expected to improve the effectiveness and efficiency of the licensing processes for future applicants.

License Renewal

The reactor license renewal program provides a stable and predictable regulatory process to implement the NRC's technical and regulatory requirements for the renewal of nuclear power plant licenses. As mandated by the Atomic Energy Act, the NRC issued original reactor operating licenses for 40 years, which may be renewed for up to an additional 20 years. The review process for renewal applications provides continued assurance that the level of safety provided by an applicant's current licensing basis will be maintained throughout the extended period of operation. To date, the NRC has received applications to renew the licenses for 49 units at 28 sites and has renewed the licenses for 35 units at 20 sites (see Figure 3). The NRC is currently reviewing applications to renew the licenses for the remaining 14 units at 8 sites. As of September 2005, the NRC had issued in FY 2005 renewed licenses for Dresden Nuclear Power Station Units 2 and 3, Farley Units 1 and 2, Arkansas Nuclear One Unit 2, and D.C. Cook Units 1 and 2.

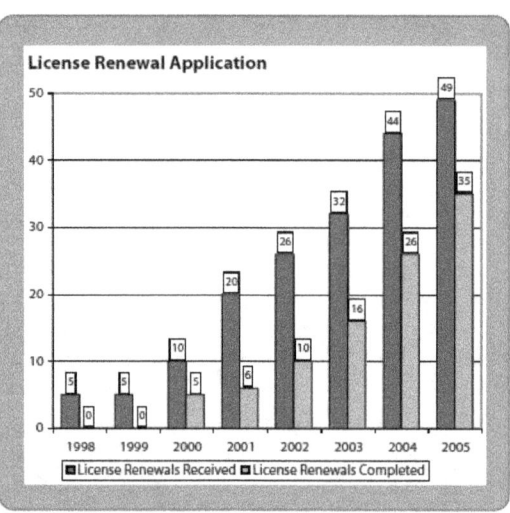

Figure 3

The NRC expects that almost all of the currently licensed units will ultimately apply to renew their licenses. To establish a stable and predictable process, the NRC has specified a timeliness goal of 22 months for those reviews that do not involve a hearing. The NRC met all established schedules for completing license renewal reviews in FY 2005.

Reactor Rulemaking

On October 1, 2004, the NRC issued a final revision to 10 CFR 50.55a, "Codes and Standards." The final rule incorporated by reference several American Society of Mechanical Engineers Boiler and Pressure Vessel Codes. These codes address construction of nuclear power plant components, inservice inspection of nuclear power plant components, and rules for the operation and maintenance of nuclear power plants.

On November 22, 2004, the NRC published a revision to 10 CFR 50.69, "Risk-Informed Categorization and Treatment of Structures, Systems and Components for Nuclear Power Reactors." The revised regulation gives nuclear power plant licensees a voluntary alternative in complying with selected deterministic requirements in the Commission's regulations. The risk-informed regulation establishes an alternate set of requirements incorporating up-to-date analytic tools and risk insights to further enhance plant safety by enabling nuclear power plant licensees to determine more precisely the safety significance of reactor systems, structures, and components and maintain these structures, systems, and components in a manner commensurate with their safety.

The agency issued a revision to 10 CFR Part 50, Appendix E, "Emergency Planning and Preparedness for Production and Utilization Facilities." The final rule amended the regulations related to the NRC approval of licensee changes to emergency action levels and established the exercise requirements for co-located licensees. The final rule was published January 26, 2005.

The NRC issued a proposed revision to 10 CFR 50.48, "Fire Protection." The proposed rule gives nuclear power plant licensees a voluntary alternative to complying with selected deterministic requirements in the Commission's regulations. The proposed rulemaking would allow licensees to rely on operator manual actions based on a performance-based and risk-informed post-fire analysis. The proposed rule was published March 7, 2005.

The NRC proposed an amendment to 10 CFR Part 52, "Early Site Permits; Standard Design Certifications; and Combined Licenses for Nuclear Power Plants." The proposed rule would amend the regulations to certify the AP1000 design. The proposed rule was published on April 18, 2005.

Reactor Licensing Homeland Security

During FY 2005, the NRC continued its efforts to enhance safety of licensees by conducting two plant-specific assessments at each nuclear power plant to identify appropriate measures that could be taken to mitigate the effects of a broad range of terrorist threats. These assessments are intended to identify further effective mitigation capabilities. The first will be an independent NRC assessment of spent fuel pools and will be completed for all plants by the end of 2005. The NRC is developing detailed plans for the second assessment that will address the reactor core and containment at each plant. These assessments are expected to be completed before the end of FY 2006.

The NRC issued safety evaluations approving all of the revised security plans for the nuclear facilities that the agency licenses. As required by agency orders, all licensees have implemented their revised security plans. The agency continues oversight of security plan implementation at all licensee facilities through the Reactor Oversight Process and the Fuel Cycle Facility Operational Safety and Safeguards Inspection Program. Insights from the security plan reviews were used to ensure that the baseline inspections were performed in a manner consistent with licensing decisions in the areas of access authorization and fitness-for-duty and fatigue process checks, testing and maintenance, security training, protective strategy evaluation, owner-controlled area patrols, and security plan changes. A pilot Significance Determination Process was conducted, and the agency inspected all baseline inspection program requirements at one plant.

As part of the agency's successful participation in the development of the Federal National Response Plan and the National Incident Management System, the agency successfully modified its plans, protocols, and procedures to fully implement the requirements of the National Response Plan and the National Incident Management System. The agency also developed an emergency preparedness and response improvement initiatives plan to upgrade the agency's response and preparedness capabilities. The agency also worked with other Federal agencies (the Federal Emergency Management Agency and the Department of Homeland Security) to upgrade the emergency response and incident preparedness capabilities of its facilities through both licensing and regulation.

International Activities

In May 2005, the NRC cooperated with the International Atomic Energy Agency and plant personnel in an Operational Safety Review Team mission at the Brunswick Nuclear Power Plant in Southport, North Carolina. A 12 member team, consisting of experts from the International Atomic Energy Agency, Canada, Slovakia, the United Kingdom, Russia, Japan, France, the Czech Republic, Germany, and Slovenia conducted an in-depth review of the safety and reliability of plant operation. Operational Safety Review Team (OSART) missions give the host countries (plant and utility management, the NRC, and other government authorities) an objective assessment of operational safety at the host plant with respect to proven international performance and practices.

The NRC has been a leader in working with other national nuclear regulatory authorities, particularly in the arena of advanced reactor oversight. In particular, the NRC has proposed an initiative, the multinational design approval program that will allow several regulatory authorities to work together to share expertise and resources in reviewing new and future reactor designs. The NRC presented this initiative to the other national nuclear regulatory authorities during a meeting at the Nuclear Energy Agency in June 2005.

The NRC headed the U.S. Government delegation to the Third National Report Review Meeting of Contracting Parties to the Convention on Nuclear Safety held April 11–22, 2005. The objective of the Convention on Nuclear Safety is to achieve and maintain a high level of nuclear safety worldwide by enhancing national programs and increasing international cooperation. The Convention is an intensive peer-review process. The U.S. delegation gained valuable insights about the status of nuclear safety in other countries. These insights will help shape future NRC and U.S. Government interactions by focusing on areas where safety can be most improved in the United States and throughout the world.

Nuclear Reactor Inspection

The NRC's Reactor Oversight Process verifies that nuclear plants are being operated safely and in accordance with the NRC's rules and regulations. The NRC has full authority to take whatever action is necessary to protect public health and safety and may demand immediate licensee action, up to and including plant shutdown. The Reactor Oversight Process uses both inspection findings and performance indicators to assess the performance of each plant within a regulatory framework of seven cornerstones of safety. Toward that end, the NRC performs a program of baseline inspections at each plant and may perform supplemental inspections and take additional actions to ensure that the plants address significant issues. The NRC communicates the results of its oversight process by posting plant-specific inspection findings and performance indicator information and industry-level indicators, on the NRC's public Web site. The NRC also conducts public meetings with licensees to discuss the results of the NRC's assessments of licensee performance.

The Reactor Oversight Process is designed to ensure protection of public health and safety and the environment and supports the agency Safety goal by focusing NRC and industry attention on risk-significant activities. The process consists of risk-informed inspections, a Significance Determination Process to evaluate the risk significance of inspection findings, licensee-reported performance indicator information, and assessment and enforcement activities.

As a second layer of assessment, the NRC trends the qualitative indicators of licensee safety performance, evaluates the indicators for adverse trends, and takes action to improve industry performance and/or to provide feedback to the NRC's regulatory oversight processes.

In FY 2005, the NRC continued to integrate improvements into its regulatory process as a result of the annual Reactor Oversight Process self-assessments and completed the 2004 assessment in April 2005. The self-assessment results indicate that the Reactor Oversight Process continues to be effective in monitoring operating nuclear power plant activities and focusing NRC resources on significant performance issues and in supporting the NRC's strategic goals. The Reactor Oversight Process has been successful in achieving the goals of being objective, risk-informed, understandable, and predictable and thereby also supports the agency's Effectiveness goal to ensure that NRC actions are effective, efficient, realistic, and timely.

During FY 2005, the NRC maintained its focus on stakeholder involvement and continual improvement of the Reactor Oversight Process as a result of stakeholder feedback and lessons learned. The agency's assessments continue to show that the Reactor Oversight Process has resulted in a more objective, risk-informed, and predictable regulatory process, which focuses NRC and licensee resources on aspects of plant performance that have the greatest impact on safe plant operation. The responses to the NRC's annual survey of external stakeholders, which solicited feedback on the Reactor Oversight Process, were generally favorable. However, some stakeholders raised concerns about the timeliness and subjectivity of the Significance Determination Process, the effectiveness of the performance indicator program, and other areas considered needing improvement. The NRC has evaluated these and other stakeholder insights and has several agency initiatives underway to address stakeholder concerns, including increased management oversight of the inspection finding assessment process.

The NRC continued to implement improvements to the Reactor Oversight Process. Examples of the NRC's improvement initiatives are the development of a Mitigating Systems Performance Index, improvements in the effectiveness of the design engineering inspections, reassessment of baseline inspection procedures to better align the available inspection resources with risk-significant areas, and improvements in the Significance Determination Process.

Davis-Besse Lessons Learned

In March 2002, FirstEnergy Nuclear Operating Company, the licensee for the Davis-Besse Nuclear Power Station, discovered a cavity in the plant's reactor pressure vessel head. The NRC inspected and assessed this safety issue; directed licensees to report the condition of their reactor pressure vessel heads, past incidents of boric acid leakage, and their inspection and examination programs; assessed the operating experience function; and chartered the Davis-Besse Lessons Learned Task Force to look for ways to improve NRC performance. Forty-nine recommendations were adopted and addressed through action plans to incorporate the reactor pressure vessel inspection requirements into the *Code of Federal Regulations*, coordinate research activities for evaluating potential improvements in detection and monitoring of leakage in reactor coolant system components, assess the NRC's Operating Experience Program, and change the NRC's inspection program. In FY 2005, the NRC implemented changes recommended by a task force that analyzed how the NRC evaluates and disseminates operating experience to NRC staff, licensees, and others. The task force determined that the NRC's current reactor operating experience activities include the necessary functions of identifying short- and long-term safety issues, assessing their significance, and taking corrective action to address the issues. The task force recommended enhancements to current activities. The NRC is currently implementing the recommendations through the new Reactor Operating Experience Program, initiated on January 1, 2005. The new program establishes a single organization to systematically collect, communicate, and evaluate operating experience information, including foreign operating experience. The new process makes significant use of information technology to consolidate a large collection of individual databases and Web sources of information onto a single Web access page and make operating experience information readily available to internal users and members of the public.

A new communication tool to promptly notify NRC staff members of new operating experience in their areas of expertise or practice has been developed and we have created teams of technical review groups to systematically and periodically assess operational data in their specialized areas to identify trends and insights for further attention.

Reactor Inspection Homeland Security

Emergency Preparedness Inspection

The agency provides oversight of inspection activities for emergency preparedness at nuclear reactor facilities as a part of its overall reactor oversight activities and provides technical support for incident response activities regarding actual incidents and exercises, as well as support for inspection activities. The agency successfully modified its plans, protocols, and procedures during the implementation of National Response Plan/ National Incident Management System. As part of this effort, the agency developed an emergency preparedness and response improvement initiatives plan, designed to enable the agency to upgrade its response and preparedness capabilities. The agency also worked with other Federal agencies (FEMA/DHS) to upgrade the Emergency Response and Incident Preparedness capabilities of its facilities through both licensing and regulation.

Security Inspection

The agency completed its transitional force-on–force inspection activities, and began full program implementation in FY 2005. The agency completed 20 force-on–force inspections and schedules each site to be inspected once every 3 years. The agency also identified and addressed challenges in improving the realism of the force-on–force inspection activities. The agency implemented enhancements such as improved Multiple Integrated Laser Engagement System gear, real-time information collection on inspection participants, and site defensibility upgrades.

Safety Research

The NRC's reactor safety research program evaluates and resolves safety issues for nuclear power plants, proposes regulatory improvements, coordinates agency activities related to consensus and voluntary standards for agency use, assesses the effectiveness of certain NRC programs, and evaluates operational events to identify precursors to accidents. The agency conducts its research programs to evaluate areas of potentially high risk or safety significance, reduce uncertainties in risk assessments, and develop the technical basis to support realistic safety decisions. Where possible, the NRC engages in cooperative research with other Government agencies (such as the Department of Energy and the National Aeronautics and Space Administration), the nuclear industry, universities, and international partners. The research program includes key activities to support the agency in addressing issues in the areas of emergency core cooling system sump performance, risk analyses and rulemaking, fire safety, fuel and thermal-hydraulic, severe accident, materials degradation and structural integrity, digital safety systems, and radiological protection.

Emergency Core Cooling System Sump Performance

The NRC established Generic Safety Issue (GSI)-191, "Assessment of Debris Accumulation on Pressurized-water Reactor Sump Performance," to determine whether the transport and accumulation of debris in pressurized-water reactor containments following a loss-of-coolant accident will impede the long-term operation of the emergency core cooling system or containment spray system. Based on events at boiling-water reactors and research findings for pressurized-water reactors, the NRC issued Generic Letter 2004-02 requesting plant-specific evaluations at pressurized-water reactors to ensure compliance with design requirements. The NRC is conducting additional research in a few areas to support these evaluation efforts and provide confirmatory information. These areas include research on chemical effects to determine if the pressurized-water reactor sump pool environment generates byproducts which contribute to sump clogging, research on pump head losses caused by accumulation of containment materials and chemical byproducts, and research on the effect of injected debris on valve performance. During FY 2005, the NRC and the Electric Power Research Institute, under a memorandum of understanding, completed research on the formation of chemical byproducts. The NRC also conducted research on pump head loss and valve performance to support the staff review of licensee responses to Generic Letter 2004-02.

Risk Analysis and Rulemaking

The reactor research program supports the agency's efforts to use risk information in all appropriate aspects of regulatory decisionmaking, applies risk assessment technology to resolve safety issues, develops a risk-informed regulatory framework, and focuses regulatory activities on significant aspects of licensed activities. In FY 2005, the NRC's risk assessment research program supported numerous agency programs, including the reactor oversight and operating experience programs, rulemaking initiatives, human reliability analysis, new reactor licensing, and risk communication. During FY 2005, the NRC continued to provide support for the industry-wide implementation of the Mitigating System Performance Index, a risk-informed performance indicator intended to address concerns with the current safety system indicators. The NRC completed the accident sequence precursor evaluation for degraded conditions initially identified at the Davis-Besse Nuclear Power Plant in 2002. During FY 2005, the NRC issued a draft report on a reevaluation of station blackout risk. In support of new reactor licensing, the NRC issued the first draft of a regulatory framework which will provide the technical basis for technology-neutral regulations for public comment. To facilitate the communication of risk information, the NRC published guidelines for effective internal communication of risk information. In the area of human reliability analysis, the NRC published a report on good practices for human reliability analysis to support risk-informed decisionmaking.

Fire Safety Research

The NRC's fire risk research program supports regulatory activities in the areas of fire protection and fire risk analysis. During FY 2005, the research program focused on activities supporting the implementation of a new risk-informed, performance-based fire protection rule. This work focused on (1) development of state-of-the-art fire performance risk assessment methodology as part of the fire risk re-quantification effort and (2) the verification and validation of fire models. On September 13, 2005, the NRC issued a joint report with the Electric Power Research Institute describing state of the art fire risk assessment methods for nuclear power facilities. To support the development of guidance documents for risk-informed, performance-based fire protection programs, the NRC has continued an international cooperative effort with the National Institute of Standards and Technology to benchmark, verify, and validate fire models. Other research activities in the fire safety area have included full-scale endurance testing of the Hemyc and MT Electrical Raceway Fire Barrier Systems, which are designed to protect certain plant equipment needed to achieve a safe-shutdown condition during a nuclear plant fire.

Fuel and Thermal-Hydraulic Research

The NRC is studying the behavior of fuel with advanced cladding at high burnup. This experimental work confirms that safety is being maintained as the industry seeks the economies of advanced fuel designs and high utilization (burnup). This work will provide the technical basis for using advanced fuel cladding alloys and will permit higher fuel burnup. This first-of-a-kind experimental program and new analytic methods, will establish new safety limits for energy deposition and clad oxidation during postulated accidents. The NRC, the international community, and industry are co-funding much of this work to achieve significant efficiencies.

The NRC has developed an independent audit capability for assessing the performance of mixed-oxide fuels under normal, transient, and accident conditions and is now assessing fuel performance. This work provides the technical basis for using and disposing of weapons-grade plutonium in a power reactor. These analyses and experiments to determine the adequacy of loss-of-coolant accident criteria for high-burnup and mixed-oxide fuels support development of performance-based fuel criteria for the 10 CFR 50.46 rulemaking and assessment of the adequacy of the revised source term (NUREG-1468) for high-burnup and mixed-oxide fuels.

The NRC has an extensive thermal-hydraulic program comprising experimental testing, model development, and validation. These models and experimental results are used in developing the technical basis for risk-informing the regulations, addressing emergent safety issues, and providing the capability for independent audit calculations for proposed new designs. This effort supports the staff review of the AP1000 and Economic Simplified Boiling-Water Reactor new reactor designs.

Severe Accident Research

The NRC has developed an independent audit capability for rare-event (severe-accident) analysis and is participating in research to maintain severe accident expertise within the NRC. The severe accident research allows the NRC to assess, develop and maintain an independent state-of-the-art severe accident analytical tool (MELCOR code) for risk-informing its regulations, assessing the security of nuclear power plants and spent fuel pools, certifying new reactor designs, and providing a technical basis for determining whether the use of mixed-oxide fuel and high-burnup uranium fuel in operating reactors will pose an undue risk to the health and safety of the public.

A recently released version (Version 1.8.6) of the MELCOR code will be used for independent severe-accident analysis to support the agency review of the Economic Simplified Boiling-Water Reactor design. Test data on air oxidation of nuclear fuel cladding was used to realistically assess the potential for a zirconium fire in a spent fuel pool accident. Realistic thermal-hydraulics and severe accident and consequence analyses were performed to assess the protection of nuclear power plants, including spent fuels pools against postulated terrorist attacks. These activities contribute to the Nuclear Reactor Safety goal to prevent radiation-related deaths and illness, promote the common defense and security, and protect the environment in the use of civilian nuclear reactors.

Materials Degradation and Structural Integrity Research

The ability of structures, systems, and components to withstand normal operational loads, design-basis loads, and accidental loads (including natural hazards such as seismic events, tornados, and floods) is important to safe operation of nuclear power plants. Recent events related to the cracking of nickel-based alloys and associated weldments (e.g., cracking of the control rod drive mechanism nozzles at pressurized-water reactors) have highlighted the importance of aging and degradation research and have focused worldwide interest on proactive management of the degradations (identifying components susceptible to degradation and taking steps to avoid, or finding and dealing with degradation before any significant loss of safety margin). The NRC conducted a research study using a panel of international experts to identify and evaluate the degradation potential for several thousand components in light-water reactors. The components with a higher-likelihood of degradation were identified for potential inclusion in proactive materials degradation management programs. The NRC is continuing to study these components to assess the need for further NRC actions. The NRC conducted a meeting in August 2005 to develop an international cooperative group and a plan for the research needed for implementing proactive materials degradation management programs. Potential participants were identified, and a detailed program plan was initiated. The NRC also evaluated the adequacy of risk assessments of passive-component degradation and the integration of the results into the regulatory decisionmaking process. Improvements were identified for implementation.

Digital Safety Systems Research

Nuclear facility licensees are replacing analog instrumentation and control equipment with digital equipment. The main reasons for these analog-to-digital upgrades are that analog replacement parts are becoming more difficult to obtain and that digital systems potentially offer better performance and more features than analog systems. There are challenges for the agency and the nuclear industry stemming from the introduction of this new technology into nuclear facilities. Several current projects are being conducted to provide the technical basis for assessing the ability of existing digital technologies to perform their intended functions under the adverse environmental conditions expected in a nuclear power plant, such as electromagnetic and radio frequency interference and abnormal conditions such as smoke and steam environments. The NRC is also conducting research to advance probabilistic risk assessment of complex digital safety systems, including software-based and commercial off-the-shelf systems. This research leverages work that has been performed for other agencies and countries to maximize the efficient use of NRC resources.

In addition, new advanced reactor plants are expected to use advanced digital instrumentation and control systems. These systems will incorporate design features that do not exist in the current generation of U.S. nuclear power plants. Several current projects are examining emerging technologies in digital systems to identify issues that must be addressed in the licensing process and provide the technical basis for the agency's safety review.

Radiological Protection Research

The NRC provided comments on new draft recommendations on radiation exposure from the International Commission on Radiological Protection, which periodically evaluates current information on radiation health effects and then revises its recommendations as appropriate. The NRC maintained some ongoing activities to monitor radiation exposures and events, including the Radiation Exposure and Information Reporting System database of occupational exposure records for certain classes of NRC licensees. NRC guidance on recording and reporting occupational radiation exposure was updated and Abnormal Occurrences in 2004 were identified and reported to Congress.

Industry Trends Program

The NRC measures the effectiveness of its Nuclear Reactor Safety program activities based on the continued safe operation of the Nation's nuclear power plants. In addition to monitoring the performance of individual plants, the NRC compiles data on overall safety performance using several industry-level performance indicators, some of which are addressed in the following pages. The NRC analyzes data that is outside of the prediction limits for safety that are set using statistical analysis. These indicators show significant improvement in the long-term trends for safety performance of nuclear power plants since 1988, the baseline year for the statistical analyses. The baseline year for the precursor occurrence rate is 1993. For ease of viewing, all the charts in this section display data since 1993.

The industry safety indicators are derived through complex engineering and scientific analyses by the NRC's Office of Nuclear Reactor Regulation and Office of Nuclear Regulatory Research. The analyses of some events for FY 2004 and FY 2005 are still ongoing. The performance indicator results are subject to minor variations as licensees submit revisions to the source data and may differ slightly from data reported in previous years as a result of refinements in data quality. The results of these analyses are reported annually to both the Commission and to Congress.

The Industry's Safety Performance Record

Several industry indicators of safety performance show significant statistical improvement since 1988. One such indicator is significant events, which meet specific criteria such as degradation of important safety equipment (see Figure 4).

The total (collective) radiation dose received by workers is an indication of the radiological challenges of maintaining and operating nuclear power plants. The trend shows a reduction in collective dose since 1988 and demonstrates the effectiveness of the controls on radiation exposure implemented to meet these challenges (see Figure 5).

Safety systems mitigate off-normal events such as the widespread power blackout in August 2003 by providing reactor core cooling and water addition. Actuations of safety systems that are monitored include certain emergency core cooling and emergency electrical power systems. Actuations can occur as a result of "false alarms" (such as testing errors) or in response to actual events. The statistical trend for number of safety system actuations highlights the improved performance since 1988 (see Figure 6).

A scram is a basic reactor protection safety function that shuts down the reactor by inserting control rods into the reactor core. Scrams can result from events that range from relatively minor incidents or human error to precursors of accidents. The massive power blackout in August 2003 accounts for most of the increase in scrams in FY 2003, but has not affected the statistical trend for number of scrams, which has been declining steadily since 1988 (see Figure 7).

The NRC assesses the risk significance of events at plants. A precursor event is an event that has a probability of greater than 1 in 1 million of leading to substantial damage to the reactor fuel. There is no statistically significant adverse trend in the occurrence rate of precursor events since 1993, the

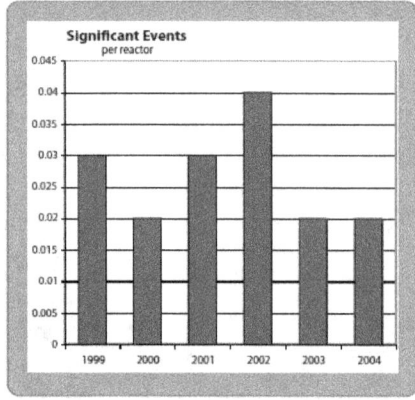

Figure 4

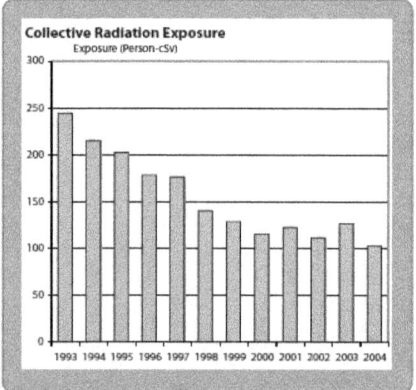

Figure 5

Figure 6

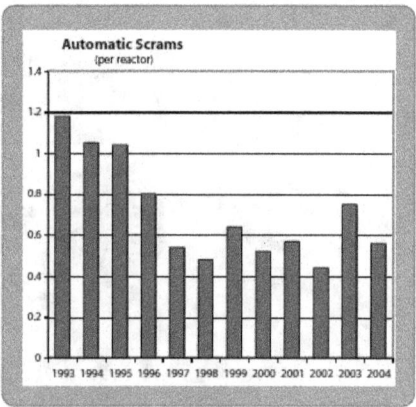

Figure 7

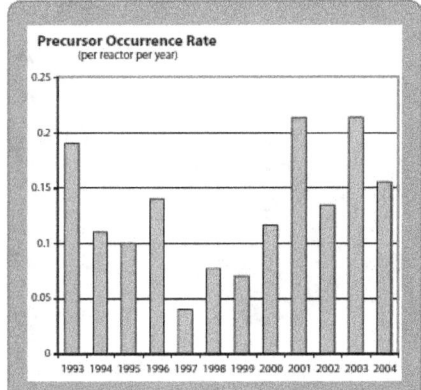

Figure 8

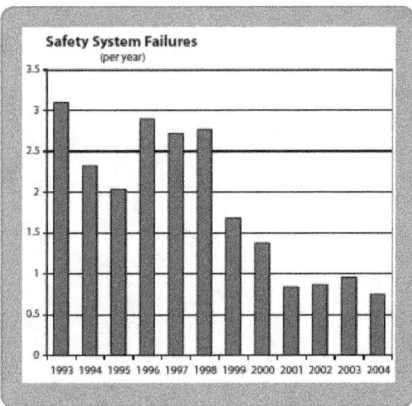

Figure 9

baseline year for the statistical analysis. Due to the complexities associated with evaluating precursor events, the data always lag behind other indicators. Precursor data through FY 2004 (FY 2004 contains preliminary data) is shown (see Figure 8).

Safety system failures include any events or conditions that could prevent a safety system from fulfilling its safety function. The statistical trend for number of safety system failures across the industry has declined since 1988 (see Figure 9).

The Power Generation and Average Capacity Factor indicators are not a part of the NRC's Industry Trends Program. The data are obtained from the Department of Energy, and are displayed from 1993 through 2004. Improvements in safety have occurred at a time when nuclear power generation has increased significantly, from 610,000 gigawatt hours in 1993 to approximately 789,000 gigawatt hours in 2004 (see Figure 10).

The average annual capacity factor, a measure of power plant efficiency, has increased from 73 percent in 1993 to 90.5 percent in 2004 (see Figure 11).

The NRC's Role in Improving Safety

The improvement in the safety performance of nuclear power plants is the result of the combined efforts of the nuclear industry and the NRC. Both the nuclear industry and the NRC have gained experience in the operation and maintenance of nuclear power facilities. The NRC establishes safety standards and safety requirements, performs in-depth technical reviews of proposed reactor designs, and oversees plant operating performance. The NRC will not allow licensees to operate their plants if safety performance falls below acceptable levels. Licensees are primarily responsible for maintaining safety. They are responsible for designing, maintaining, and operating nuclear power plants in a manner that provides adequate protection of public health and safety.

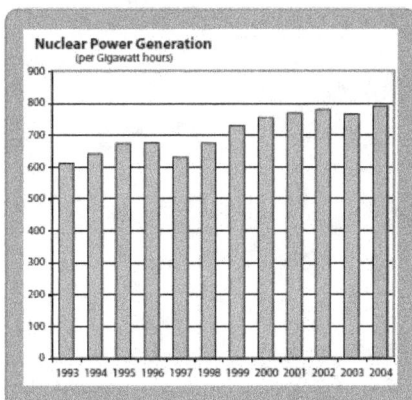

Figure 10

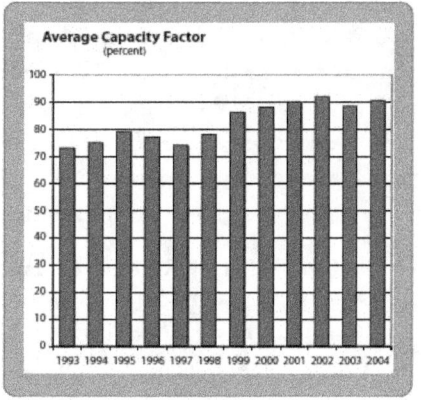

Figure 11

Plant operating experience data have yielded a steady stream of improvements in the reliability of plant systems and components, plant operating procedures, training of power plant operators, and regulatory oversight.

Funding for Achieving Goals

The Nuclear Reactor Safety budget was $443.1 million for FY 2005. This budget was allocated to two major activities, reactor licensing and reactor inspection (see Figure 12).

Each activity has a specific and linked role in ensuring safety at nuclear power plants. The licensing activity sets the standards and procedures for operating nuclear power plants, and the inspection and performance assessment activity results in inspection of the plants and the collection of information to ensure that licensing obligations are being met and that each plant's performance is within the required safety range.

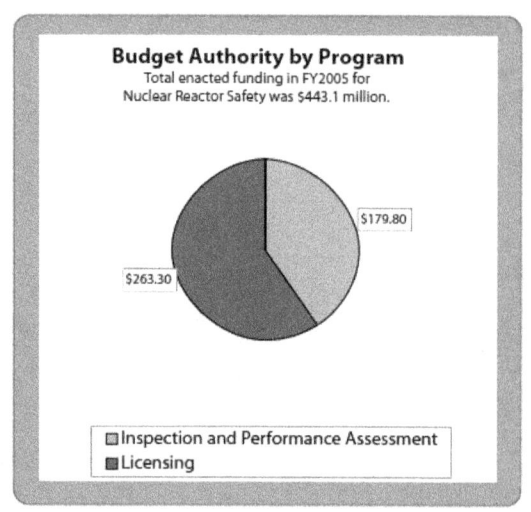

Figure 12

Program Evaluation

In FY 2005, the NRC continued to integrate improvements into its regulatory process as a result of the annual Reactor Oversight Process self-assessments. The NRC completed the 2004 assessment in April 2005. The report, "Reactor Oversight Process Self-Assessment for Calendar Year 2004," (SECY-05-0070), is available through the NRC public Web site. The 2004 self-assessment results indicate that the Reactor Oversight Process met its program goals of being objective, risk-informed, understandable, and predictable. The Reactor Oversight Process was also effective in supporting the NRC's strategic goals of Safety, Security, Openness, Effectiveness, and Management Excellence. The NRC implemented several additional Reactor Oversight Process improvements recommended by the Davis-Besse Lessons Learned Task Force, the Office of the Inspector General, other independent evaluations, and internal and external stakeholders. The Reactor Oversight Process self-assessment metrics were all met, except for one of the eight performance indicator metrics, four of nine Significance Determination Process metrics, one of eleven assessment metrics and two of eighteen overall metrics.

Although significant progress has been made in 2004, the NRC expects to make continued improvements to the Reactor Oversight Process based on lessons learned and stakeholder feedback. The NRC will continue to actively solicit input from the NRC's internal and external stakeholders and will evaluate potential program improvements via the ongoing self-assessment process. The NRC will also continue to report the results of its annual self-assessment as part of the Commission briefing following the Agency Action Review Meeting.

The NRC has started to implement many of the recommendations developed during the FY 2004 program evaluation of the agency's operating experience program. On January 1, 2005, the NRC implemented many of these changes through the initiation of the enhanced power reactor Operating Experience Program. The NRC's Office of the Inspector General conducted an independent review of the NRC's Reactor Operating Experience Task Force Report (Audit Report OIG-04-A-13). The Office of the Inspector General concluded that the Reactor Operating Experience Task Force Report was comprehensive and that the report's conclusions and recommendations adequately addressed the identified program weaknesses. The Inspector General also documented six recommendations to further enhance the effectiveness of the Operating Experience Program. In FY 2005, the NRC resolved four recommendations, and two recommendations remain open pending Inspector General review of program documents scheduled to be revised in FY 2006.

In addition, a task force performed an assessment of the NRC's process for reviewing the scoping and screening portions of license renewal applications to verify compliance with the requirements of 10 CFR Part 54, "Requirements for Renewal of Operating Licenses for Nuclear Power Plants." That assessment included a review and audit of the applicant's scoping and screening methodology, a technical review of the scoping and screening results documented in the application, and inspection of the implementation of the scoping and screening results. The intent of the assessment was to determine whether the NRC can better define the interface between organizations to minimize overlapping activities, if any, and to improve the effectiveness and efficiency of the review process. The task force's assessment showed that the various NRC organizations were conducting their related activities with approved program procedures and in accordance with regulatory requirements. The team identified areas for improving the coordination and communication of activities. The NRC is currently evaluating possible approaches for implementing the team's recommendations.

In FY 2005, the Inspector General completed an audit of the NRC's Baseline Inspection Program (OIG-05-A-06, December 22, 2004). Overall, the audit found that NRC staff, licensees, and stakeholders view the Reactor Oversight Process. The Inspector General audit identified several weaknesses in the baseline program, and the report made 10 recommendations to improve the efficiency and effectiveness of the baseline inspection program. The NRC is in the process of making program changes to address the recommendations from the audit.

In addition, a task force evaluated the NRC's program for handling licensing actions submitted by licensees for operating nuclear power plants (excluding license renewal). Such applications include proposed changes to facility operating licenses and technical specifications, requests for relief from the requirements of the American Society of Mechanical Engineers Boiler and Pressure Vessel Code, responses to orders, and proposed exemptions from NRC regulations. The team assessed the current work processes, performance measures, and data related to the timeliness and productivity of the NRC staff. The assessment also included feedback from licensees and other stakeholders as well as insights from a recent evaluation performed using the Program Assessment Rating Tool developed by the Office of Management and Budget. The team concluded that the existing program has been successful in enabling the safe operation of commercial nuclear power plants. The team identified and documented possible improvements in the management of the program, related work processes, and the performance measures. The NRC is currently evaluating the team's recommendations.

Program Assessment Rating Tool

Over the past several years, the Office of Management and Budget has conducted reviews utilizing the Program Assessment Rating Tool of the Nuclear Reactor Safety activity. The following sections include a description of the activities, the Office of Management and Budget recommendations, and NRC's response and their impact upon program performance.

Reactor Licensing

In FY 2005, the Office of Management and Budget rated the reactor licensing activity as "moderately effective," which is the second highest rating category, and gave the activity an overall score of 74. The NRC is currently awaiting the recommendations from the Office of Management and Budget, and NRC's actions and results realized will be included in next year's report.

Reactor Inspection

The Office of Management and Budget rated this activity as "effective" with an overall score of 89 in FY 2003. The activity earned high scores for Program Purpose and Design and for Program Management. It was noted that the purpose of the activity was clear, well-designed, and results-oriented. Also noted was that this activity has met all of its performance measures since Government Performance and Results Act reporting began in 1997.

The Office of Management and Budget recommended including better linkage of budget requests to NRC's annual and long-term goals and the linkage of performance measures in the organization's operating plan to support the safety performance measures in the FY 2004–FY 2009 Strategic Plan. The second recommendation was for more transparency in how allocation decisions are made and how the activity contributes to achievement of the agency's long-term goals as well as conduct a complete review of operating plan format and content to improve their effectiveness as management tools.

The NRC has responded to the first recommendation through its initiative to define outcomes and outputs that align with performance measures. Additionally, the NRC is working to improve its cost management capabilities to better align its costs with outcomes. The NRC also demonstrated via direct linkage of FY 2005 Operations Plan performance measures to the FY 2004–FY 2009 Strategic Plan strategies for meeting the Strategic Plan objective and goals. Each of the operating plan's safety performance measures reference one or more of the Strategic Plan strategies under the safety goal.

To respond to the second recommendation, the NRC has moved to the implementation of costing to the NRC's safety and security goals in the Strategic Plan beginning with the FY 2006 request. In addition, the NRC has demonstrated better linkage of budget requests to agency goals through utilization of the common prioritization process for establishing the linkage between operational activities, including the resources allocated to support these activities, and the agency's strategic and long-term goals. The NRC's Reactor Inspection and Performance Assessment program managers have responded to the Office of Management and Budget recommendation by linking operational activities and the agency's strategic and long-term goals in the revised operating plans.

NUCLEAR MATERIALS AND WASTE SAFETY

Overview

The Nuclear Materials and Waste Safety program encompasses regulatory oversight for five activities—Fuel Facilities Licensing and Inspection, Nuclear Materials Users Licensing and Inspection, High-Level Waste Repository, Decommissioning and Low-Level Waste, and Spent Fuel Storage and Transportation Licensing and Inspection. This oversight includes all regulatory activities carried out by the NRC and the Agreement States to ensure that nuclear materials and waste facilities are used in a manner that protects the public health and safety and the environment, while also protecting against radiological sabotage and theft or diversion of special nuclear materials. The following sections discuss the NRC's achievements in each of these activities.

Fuel Facilities Licensing and Inspection Activity

The Fuel Facilities Licensing and Inspection activity oversees uranium extraction, conversion, and enrichment activities and nuclear fuel fabrication facilities. The NRC licenses and inspects all commercial nuclear fuel facilities that process and fabricate uranium ore into reactor fuel. Licensing and inspection actions are a key aspect of the agency's nuclear fuel cycle safety and safeguards program. Inspection actions include detailed health, safety, safeguards, and environmental licensing reviews and inspections of licensees' programs, procedures, operations, and facilities to ensure safe and secure operations.

Each of the Nation's 37 fuel cycle facilities holds a license or certificate that specifies the materials the licensee may possess and sets restrictions on how those materials may be used. In addition to authorizing the possession and use of source, special nuclear, and byproduct material, each license or certificate establishes related licensee responsibilities (such as worker protection, environmental controls, and financial assurance). The NRC issues these fuel cycle facility licenses or certificates in accordance with requirements promulgated in the *Code of Federal Regulations*. Applications for licenses or certificates demonstrate how the licensees will operate their facilities to ensure adequate safety and safeguards.

The NRC completed 95 fuel cycle licensing actions and conducted 99 inspections, covering 204 inspection modules, at fuel cycle licensees during FY 2005 (see Figure 13). In FY 2005, the NRC began tracking fuel cycle inspection modules completed rather than inspections conducted because inspection modules focus on the specific areas being inspected (e.g., chemical, nuclear criticality safety) rather than on site visits. Therefore, tracking inspection modules completed is a better measure of program performance than the number of inspections. Because multiple modules may be completed during a given inspection activity, the number of modules will be consistently greater than the number of inspections. Beginning in FY 2006, NRC will no longer report on number of inspections completed, but will report on the number of inspection modules completed.

Figure 13

The NRC is currently involved in several significant fuel cycle licensing reviews and has recently completed several of them. Pursuant to a bilateral agreement between the Department of Energy and the Russian Federation, Duke, Cogema, and Stone & Webster submitted a request to the NRC for authorization to construct a mixed-oxide fuel fabrication facility on the Department of Energy's Savannah River site near Aiken, South Carolina. Under this agreement, the U.S. and the Russian Federation would each convert 34 metric tons of weapons-grade plutonium that has been declared excess to national security needs into forms less usable in nuclear weapons. The NRC issued a final environmental impact statement in February 2005, and on March 30, 2005, the NRC issued a construction authorization and published a final safety evaluation report on Duke, Cogema, and Stone & Webster construction authorization request. These were significant milestones in the Department of Energy's Surplus Plutonium Disposition Program. No further NRC action is required prior to construction of the mixed-oxide fuel fabrication facility.

In June 2005, the NRC completed its review of the Louisiana Energy Services license application for the National Enrichment Facility, a proposed commercial gas centrifuge uranium enrichment facility to be located in Lea County, New Mexico. The NRC's safety evaluation report (NUREG-1827) and final environmental impact statement (NUREG-1790) were issued on June 15, 2005. The NRC completed these reviews on an aggressive 18-month schedule. During the reviews, the NRC conducted three public meetings near the proposed facility to provide information on the NRC licensing process and to seek input from the public for an environmental impact statement. In preparing the final environmental impact statement, the NRC addressed nearly 4,200 comments received on the draft environmental impact statement.

USEC, Inc., submitted a license application to the NRC on August 23, 2004, for the American Centrifuge Plant, a proposed commercial gas centrifuge uranium enrichment facility to be located in Piketon, Ohio. The NRC is currently reviewing this license application. The NRC has conducted two public meetings near the proposed facility to provide information on the NRC licensing process and to seek input from the public for the environmental impact statement.

The NRC is conducting integrated safety analysis summary reviews for individual license amendment requests. These independent reviews are part of the agency's implementation of the revised regulation established in Part 70 of Title 10, of the *Code of Federal Regulations* (10 CFR Part 70), which increases the use of risk information for fuel cycle facilities. During this fiscal year, the NRC continued reviews of an integrated safety analysis submitted by BWX Technologies, Inc., and partial integrated safety analyses submitted by Westinghouse Electric Co., LLC, and Global Nuclear Fuel-Americas, LLC.

The NRC also initiated reviews of integrated safety analyses submitted by Nuclear Fuel Services, Inc., and Framatone ANP-Richland. In addition, reviews were initiated on supplemental portions of analyses submitted this fiscal year by Westinghouse Electric Co., LLC, and Global Nuclear Fuels.

For other fuel facilities, significant activities in FY 2005 include the Atomic Safety and Licensing Board panel decision of March 28, 2005, which upheld the NRC's issuance of three amendments to the Nuclear Fuel Services, Inc., license for the Blended Low Enriched Uranium Project. This project is part of a Department of Energy initiative to reduce existing supplies of surplus highly enriched uranium through reuse or disposal. Nuclear Fuel Services, Inc., has contracted with Framatome ANP, Inc., to downblend surplus highly enriched uranium into a low-enriched uranium dioxide product that will be converted to commercial reactor fuel for use in a Tennessee Valley Authority nuclear power reactor.

Regarding the NRC oversight of uranium recovery activities, in FY 2005, after successful reclamation, the Petrotomics Company and the Sohio Western Mining Company transferred ownership of the Shirley Basin South and L-Bar uranium mill tailings sites, respectively, to the U.S. Department of Energy for long-term custody, pursuant to Title II of the Uranium Mill Tailings Radiation Control Act of 1978 and the NRC's implementing regulations in 10 CFR Part 40. Subsequent to this transfer, the NRC accepted the U.S. Department of Energy's long-term surveillance plans for these sites. This acceptance established the U.S. Department of Energy as the long-term custodian and caretaker of the Shirley Basin South and L-Bar sites. In a concurrent action, the NRC terminated Petrotomics' and the Sohio Western Mining Company's specific licenses for these sites.

The agency reviews and approves facility specific physical security plans and fundamental nuclear material control and accounting plans for facilities that the agency regulates. These plans document the safeguards measures in place to deter and protect against threats of radiological sabotage and theft or diversion of special nuclear material at designated fuel cycle facilities.

This activity also supports the US-Russian effort to reduce the stockpile of weapons-grade plutonium by performing technical reviews of the exemption requests and the revised physical security plan to allow the use of four mixed oxide fuel lead test assemblies at the Catawba Nuclear Station. The purpose of the lead test assemblies effort at Catawba is to confirm that the mixed oxide fuel performs as expected in a nuclear power reactor. In FY 2005, the agency completed 12 material control and accounting and 8 physical security licensing actions, submitted the final safeguards evaluation report (NUREG-1827) for the Louisiana Energy Services National Enrichment Facility license application, completed initial reviews for the USEC American Centrifuge Plant license application and issued requests for additional information, and completed 63 export license reviews. The agency also supported material control and accountability reviews for the hearings on Duke Power's mixed oxide fuel applications.

Materials Users Licensing and Inspection Activity

The Nuclear Materials Users Licensing and Inspection activity oversees large and small users of nuclear material for industrial, medical, or academic purposes (radiographers, hospitals, private physicians, nuclear gauge users, large and small universities, and others). The NRC and 33 Agreement States regulate more than 20,000 specific and 150,000 general materials licensees. The NRC currently regulates and inspects approximately 4,500 specific licensees for the use of nuclear byproduct and other radioactive materials.

These uses include medical diagnosis and therapy, medical and biological research, academic training and research, industrial gauging and nondestructive testing, production of radiopharmaceuticals, and fabrication of commercial products (such as smoke detectors) and other radioactive sealed sources and devices. Detailed health and safety reviews and inspections of licensee procedures and facilities provide reasonable assurance of safe operations and the development of safe products. The NRC routinely inspects materials licensees to ensure that they are using nuclear materials in a safe manner, maintaining accountability of those materials, and protecting public health and safety. The NRC also analyzes operational experience from NRC and Agreement State licensees. In particular, the NRC meets regularly to evaluate the safety significance of the events reported by licensees and Agreement States.

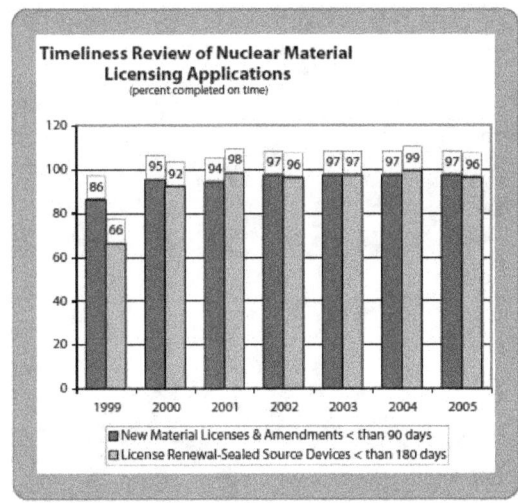

Figure 14

In FY 2005, the NRC completed review of 3,274 materials licensing actions and approximately 1,300 materials program inspections. The NRC's timeliness in reviewing nuclear material license renewals and sealed source and device designs has improved from 1999 through 2005 (see Figure 14). In FY 2005, 608 of 632 (96%) renewals and sealed source and device design reviews were completed within 180 days, and 2,568 of 2,641 (97%) of new applications and license amendments were completed within 90 days.

Under its authority for regulating nuclear material used for medical purposes, in March 2005 the NRC issued a final rule to amend the agency's requirements for training and experience in 10 CFR Part 35, "Medical Use of Byproduct Material." This rule amended the regulations for recognition of certain specialty boards whose certification may be used to demonstrate the adequacy of the training and experience of individuals to serve as medical physicists, nuclear pharmacists, radiation safety officers, and authorized users (physicians). The rule provides a more flexible and performance-based approach to the requirements, thus reducing regulatory burden. The associated guidance document has been revised to reflect the amended training and experience requirements.

The NRC worked with the Department of Energy to recover unwanted or orphaned greater-than-Class C radioactive sources that were initially identified for accelerated recovery under the Department of Energy Offsite Source Recovery Program. From the inception of the Offsite Source Recovery Program in 1997 through FY 2005, over 11,000 sources have been recovered from over 420 sites on the priority list. Several large devices were recovered for which the NRC issued a certificate of compliance to allow transport of the devices and to facilitate storage by the Department of Energy. In addition to these efforts, in FY 2005 the NRC entered into a cooperative agreement with the Conference of Radiation Control Program Directors' National Orphan Radioactive Material Disposition Program to facilitate disposition of orphaned or unwanted material held by the Agreement States or the NRC licensees.

In collaboration with the Department of Homeland Security, the Department of Energy, and other agencies, the NRC continued to assess the potential use of radioactive sources in radiological dispersion devices and to identify necessary enhancements in the control of radioactive sources. The NRC issued a proposed rule that would establish the regulatory foundation for the National Source Tracking System, a database for tracking radioactive sources of concern. The proposed rule would require the NRC and Agreement State licensees to report transactions involving the manufacture, transfer, receipt, and disposal of nationally tracked sources (Category 1 and 2 sources from the IAEA Code of Conduct). A source registry has been implemented and an interim database developed as a first stage for a truly national source tracking system. The NRC works with the Agreement States to inspect the higher priority licensees and to develop appropriate security enhancements for lower priority licensees. Final enhanced security measures will be issued and an inspection program will be implemented to verify the implementation of these measures. The agency is developing a process to screen new license applications for applicants that need to implement the enhanced security measures and to identify suspicious uses of nuclear materials.

The NRC has taken the lead in implementing portions of the International Atomic Energy Agency's Code of Conduct for the Security of Radioactive Sources. The NRC also has enhanced the security requirements for licensees that hold radioactive materials designated "radionuclides of concern in quantities of concern." The NRC continued its participation in the International Atomic Energy Agency Radiation Safety Standards Committee and its Transportation Safety Standards Committee. The NRC's involvement in these committees enhances public safety and contributes to international and domestic regulatory stability. The NRC also participated in an International Atomic Energy Agency safety standards committee for reviewing and developing of safety standards and guides for storage of spent nuclear fuel. The NRC also participated in Nuclear Energy Agency committee efforts to develop guidelines for spent nuclear fuel interim storage, and to improve radiation protection regulatory programs.

In July 2005, the NRC participated in an initial meeting with the European Commission in Luxembourg. At this meeting, substantive discussions were held on nuclear materials issues including waste, clearance, radionuclides of concern, export/import, and source tracking. This initial interaction will facilitate future coordination and cooperation with the European Commission.

State and Tribal Programs

The NRC establishes and maintains effective communications and working relationships with States, local governments, Indian tribes, and interstate organizations. The NRC has relinquished its regulatory responsibilities through agreements with 33 Agreement States in accordance with Section 274b of the Atomic Energy Act. To ensure adequate protection of public health and safety and the compatibility of Agreement State programs with the NRC programs, the NRC conducted eight Integrated Materials Performance Evaluation Program reviews of Agreement State programs. The Integrated Materials Performance Evaluation Program uses a common evaluation process that applies to both Agreement State and the NRC regional materials programs to maintain a uniform materials safety policy throughout the Nation. In addition, the NRC conducted one review of an NRC regional office and a review of the NRC Sealed Source and Device Evaluation program. Also in accordance with Section 274i of the Atomic Energy Act, the NRC modified four of the nine agreements with States to conduct additional security inspections for the NRC.

High-Level Waste Repository Activity

The High-Level Waste activity is focused on the permanent storage and disposal of high-level nuclear waste. The NRC conducts its high-level waste program in accordance with the Nuclear Waste Policy Act, as amended, and the Energy Policy Act of 1992. This legislation specifies an integrated approach and a long-range plan for high-level waste storage, transportation, and disposal. It also prescribes the roles of the NRC, the Department of Energy, and the Environmental Protection Agency with respect to the high-level waste program.

The Department of Energy is responsible for disposing of the Nation's high-level waste, beginning with site characterization and repository design, and the development, operation, and ultimate closure of a deep geologic repository. It is also responsible for characterizing the potential site at Yucca Mountain in the State of Nevada. The Environmental Protection Agency has been charged with developing environmental standards for the Yucca Mountain repository consistent with recommendations of the National Academy of Sciences.

The NRC's responsibilities include licensing decisions and regulatory oversight of the permanent storage and disposal of high-level nuclear waste. The NRC has developed and will modify as necessary technical criteria for licensing, consistent with the standards promulgated by the Environmental Protection Agency. The NRC also has extensive pre-licensing responsibilities and will issue a license after determining whether the license application that the Department of Energy ultimately submits for a geologic repository at Yucca Mountain complies with the applicable regulatory standards.

The Environmental Protection Agency and the NRC issued their standards in 2001. On July 9, 2004, both sets of standards were vacated by a Federal Court of Appeals insofar as the standards incorporated the Environmental Protection Agency 10,000-year compliance period. The Environmental Protection Agency has revised its Yucca Mountain standards to be consistent with the court decision. The NRC published a draft regulation in September 2005 to amend its Yucca Mountain regulations to reflect the new Environmental Protection Agency standards. The NRC staff has also been conducting efforts to modify computer codes and run calculations to project more than 10,000 years into the future. Early in FY 2005, the Department of Energy determined that it needed more time to prepare its license application and that they would not meet their projected December 2004 submission date.

The NRC continued to focus on the actions needed to lay the groundwork for the NRC to independently conduct a license application review during FY 2005. The NRC has been working with the Department of Energy to address key technical issues and raise issues that could impact the quality of the license application. The NRC has 293 agreements with Department of Energy related to nine key technical issues. These agreements were developed to incorporate sound science into the review of the Yucca Mountain license application. Using the risk insights report to focus pre-licensing activities on significant risk issues, the NRC completed an evaluation of high-risk agreements by the end of the 1st quarter of FY 2005 and finished evaluating moderate to low-ranked agreements by the end of the 2nd quarter.

In FY 2005, the NRC issued an update of the Consolidated Issue Resolution Status Report (NUREG-1762, Rev. 1, April 2005). This publicly available report summarizes the status of technical information developed in the course of pre-licensing interactions between the NRC and the Department of Energy. The report covers issues related to pre-closure safety, post-closure performance, and other aspects of the proposed repository. The NRC expects to use the revised report and the Yucca Mountain Review Plan to conduct a risk-informed review of a license application.

The NRC enhanced its electronic information exchange capability to enable the electronic receipt of high level waste documentary material. The electronic hearing docket was used in the proceeding for the Pre-License Application Presiding Officer. The NRC obtained security approval to deploy the protective order file to support the proceeding. The NRC tested its preparedness "end-to-end" exercises on how organizations processes, procedures, functions, and systems receive, process, and respond to documents and filings. The agency's management group completed the operational readiness review for the release and determined that it met the service-level requirements and functionality for the pre-license application phase.

The NRC is investing in a digital data management system that will provide the necessary technology and functionality to meet the agency's obligation to conduct the adjudicatory proceeding for the high-level waste repository. The digital data management system will provide information technology and audio/visual capabilities in at least two hearing rooms (one in the Las Vegas area near Yucca Mountain site, the other at NRC headquarters in Rockville, Maryland); enable the creation and use of an integrated, comprehensive digital record for the high level waste repository licensing proceeding; record, store, and display the text and image of documents presented in the hearing; permit access and retrieval of the entire record; allow counsel for the parties to electronically bring prepared materials to the evidentiary hearing; and provide continual real-time access to the hearing record by the presiding officer and distribution to the parties in the litigation.

Decommissioning and Low-Level Waste Activity

The Decommissioning and Low-Level Waste activity involves licensing and inspection activities at 18 decommissioning power reactors, 17 research and test reactors, 12 uranium recovery sites, as well as 40 complex materials and fuel facility sites. Decommissioning removes radioactive contamination from buildings, equipment, groundwater, and soil to levels that permit the release of the property with or without restrictions on its future use. The NRC terminates the license for decommissioned facilities after the licensees demonstrate that the residual onsite radioactivity is within the regulatory limits and sufficiently low to protect the health and safety of the public and the environment. The criteria for terminating a license are defined in Subpart E of 10 CFR Part 20.

During FY 2005, the NRC oversaw decommissioning activities at numerous complex sites and power reactor sites. Six complex materials licenses, two uranium mill licenses, and two operating reactor license were terminated. In addition, the NRC approved the license termination plans for the Big Rock Point and Yankee Rowe power reactor sites. The NRC's review of the license termination plans, an intermediate step leading to license termination, ensures that the procedures and practices proposed by the site operators will protect the public health and safety and that the proposed decommissioning activities will make the sites suitable for release from regulatory control. Approval of a plan allows the site operator to begin the final stage of cleanup before requesting termination of the site license. Before a site license is terminated or modified, the site must be in compliance with the NRC's decommissioning criteria in 10 CFR Part 20 Subpart E. Completion of the decommissioning activities at these sites allows the sites to be returned to productive use while ensuring that residual radioactivity at the sites does not pose an unacceptable risk to the public.

During FY 2005, the NRC continued to improve the NRC's oversight of decommissioning of nuclear facilities by implementation of the Integrated Decommissioning Improvement Plan. The activities of the Integrated Decommissioning Improvement Plan will ensure that sites are decommissioned using realistic risk-informed approaches and will result in updated decommissioning guidance and new regulations to prevent problematic sites.

Under the Low-Level Radioactive Waste Policy Amendments Act of 1985, the NRC is responsible for licensing a commercial greater-than-Class disposal facility developed by the Department of Energy. There is currently no disposal facility in the U.S. for greater-than-Class wastes, and all of the waste must be temporarily stored. During FY 2005, the NRC supported the Department of Energy in its efforts to develop a disposal option for greater-than-Class-C low-level radioactive waste. The NRC provided information to the Department of Energy on inventories of greater-than-Class-C waste in the U.S. The NRC's actions will help ensure that there is a safe, secure disposal path for these wastes and that the national policy of permanent disposal of all radioactive wastes is fulfilled.

The NRC also supported the National Academies and the Government Accountability Office in studying low level waste. The NRC is funding a study of "Improving the Regulation and Management of Low-Activity Radioactive Wastes" by the National Academies and is providing information to its study committee on the NRC's activities in this area. The staff expects that the recommendations in the final report of the study committee will be useful in identifying options for effectively disposing of waste. The Government Accountability Office is studying the NRC's security measures for storing of Class B, C, and greater-than-class-C wastes. The findings and recommendations of the Government Accounting Office report will provide insights into the requirements and practices for ensuring and improving the safe storage of these wastes.

The National Defense Authorization Act of 2004 includes new NRC responsibilities for reviewing Department of Energy waste incidental to reprocessing determinations for the Savannah River Site and the Idaho National Engineering and Environmental Laboratory. Waste incidental to reprocessing is residual waste contained in tanks at the Department of Energy sites that may, in some instances, be safely grouted in place, rather than removed and disposed of in a geologic repository for high-level waste. The act requires that the Department of Energy consult with the NRC on waste incidental to reprocessing determinations and plans for disposal of waste that exceeds Class C concentrations. Additionally, the NRC is to monitor the Department of Energy compliance with the requirements of 10 CFR Part 61 and report to Congress, the State, and the Department of Energy if the NRC finds the Department of Energy is not in compliance. The NRC review helps to ensure the safe disposal of such material. In FY 2005, the NRC has initiated this work under a reimbursable agreement with the Department of Energy. Beginning in FY 2006, the NRC is authorized to conduct this work using budgeted resources.

In FY 2005, the NRC participated in the IAEA's Symposium on Low-Level Radioactive Waste Disposal to present information on United States' effort to provide for safe, efficient disposal of materials of this type. In addition, the NRC chaired a meeting on decommissioning issues, co-sponsored by the International Atomic Energy Agency and the Nuclear Energy Agency, which focused on the importance of coordinating the activities of these agencies in tracking multi-disciplinary issues and achieving a more realistic and streamlined approach to decommissioning.

Spent Fuel Storage and Transportation Activity

The Spent Fuel Storage and Transportation activities address the review and approval of Type B and fissile radioactive material transportation packages, the review and approval of dry spent fuel storage casks, and the licensing and inspection of independent spent fuel storage facilities. Millions of shipments of radioactive materials are safely and securely transported each year within the United States. Several Federal agencies share responsibility for regulating the safety and security of those shipments. The NRC closely coordinates its transportation-related activities with those of the Department of Transportation and, as appropriate, the Department of Energy. To carry out its regulatory responsibilities for spent fuel storage and radioactive material transportation, the NRC certifies and inspects both transport container package designs and spent fuel storage cask designs. The NRC also licenses and inspects the interim storage of spent fuel at both reactor sites and away-from-reactor sites. This helps to ensure that licensees provide safe interim storage of spent reactor fuel and transport nuclear materials in packages that provide a high degree of safety.

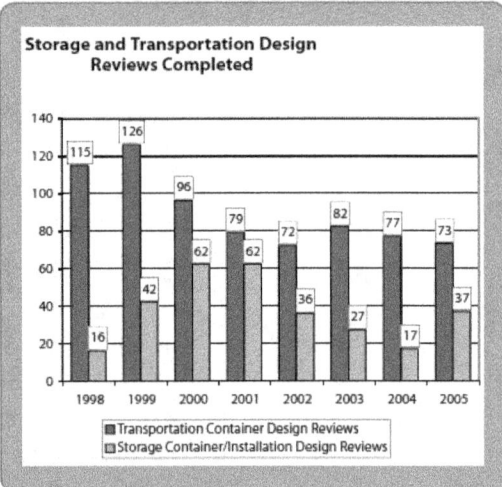

Figure 15

During 2005, the NRC completed 73 transport container design reviews and 37 storage container and installation design reviews (see Figure 15). The fluctuations in the number of transportation and storage/installation completions each year are based on licensees' needs, such as spent fuel storage capacity. The NRC's timely and effective review of transportation and interim storage licensing requests provides for the public health and safety by ensuring that shipments are made in NRC-approved packages that meet rigorous performance requirements and that spent fuel is safely stored, and thereby enabling continued reactor operations. During 2005, the NRC also conducted 21 inspections of independent spent fuel installations, and radioactive material package certificate holders.

The NRC devoted significant effort to the Private Fuel Storage license application to construct and operate an away-from-reactor independent spent fuel storage installation on the reservation of the Skull Valley Band of Goshute Indians, a Federally recognized Indian tribe. The Atomic Safety and Licensing Board Panel completed hearings on the consequences of a military aircraft crash in mid-September 2004. The Atomic Safety and Licensing Board panel issued their decision in February 2005 and issued a decision in

May 2005 on an appeal of their February decision. The Commission recently authorized the NRC staff to assist the Private Fuel storage license once the staff has made the requisite findings under NRC regulations.

In the past year the NRC issued new independent spent fuel storage installation licenses to the Department of Energy for the Idaho Spent Fuel facility, and the Diablo Canyon nuclear power plant. In addition, licenses were renewed for the H.B. Robinson, G.E. Morris, and Surry independent spent fuel storage installation. These were the first ever independent spent fuel storage installation license renewals to be issued for a 40-year period. These licensing actions will provide for the safe storage of spent fuel while allowing continued licensee operations.

The National Academy of Sciences delivered a classified report on spent fuel transportation security to the House and Senate Committees on Appropriations in July 2004 and published an unclassified summary in March 2005. The NRC responded to Congress with a report on March 14, 2005, describing the specific actions the NRC has taken in response to the National Academy of Sciences recommendations.

The agency finalized its Radioactive Material Quantities of Concern and Additional Security Measures on April 26, 2005 and continues to cooperate with the Department of Homeland Security and the Department of Transportation to enhance security for transported radioactive materials.

Materials and Waste Safety Research

The NRC has undertaken research activities to develop risk assessment tools, methods, and guidance for implementing risk-informed approaches for materials applications. The ultimate goal of these activities is to develop a technical basis to risk-inform the regulatory requirements for materials licenses. The need for more realistic tools for accurately assessing radiation doses to workers and the public is also being addressed. In addition, research activities are being undertaken to develop information on the currently licensed sources and materials that will support a rulemaking to risk-inform the regulations for using byproduct and source material.

The NRC is also conducting research for the development of a human reliability analysis capability specific to the materials program to help reduce the misuse of radioisotopes and radioactive material in medical and industrial applications and to develop tools to improve the NRC reviews of spent fuel handling.

Research activities support a number of the NRC's nuclear waste activities. One ongoing research study is to develop information and tools to assess the movement of radionuclides in the environment resulting from decommissioning and waste management activities. Another study concerns dose to the public from these activities.

The NRC published three reports intended to improve decommissioning reviews during this fiscal year. The first was NUREG/CP-0187, "Proceedings of the International Workshop on Uncertainty, Sensitivity, and Parameter Estimation for Multimedia Environmental Modeling," October 2004. The report concerns better ways to evaluate uncertainty in assessing environmental systems performance. The second was NUREG/CR-6870, "Consideration of Geochemical Issues in Groundwater Restoration at Uranium In Situ Leach Mining Facilities," June 2005. This report addresses how to estimate remediation costs for in-situ leach mining facilities and will directly assist licensing staff in evaluating financial assurance requirements for in situ leach mine licensees. The third report was NUREG/CR-6871, "Documentation and Applications of the Reactive Geochemical Transport Model," June 2005. This work demonstrated the application of an improved approach to modeling complex soil chemistry.

The NRC implemented several improvements in dose modeling capability to improve the agency's ability to estimate more realistically the potential long-term impact of radionuclides in the environment and enhance the agency's decisionmaking in terminating licenses.

Enhancements to three modeling or analysis tools should improve staff capabilities to evaluate sites for release. The first enhancement was a probabilistic version of RESRAD-OFFSITE. It was released for beta testing along with a draft user's manual in October 2004. This is the first version of the widely used RESRAD family of codes that can be used in cases where contamination has migrated away from the initial point of release. A second code, the FRAMES2 modeling platform with linkage to other modeling codes for specific environmental pathways was released in March 2005. This tool will be helpful in addressing sites with complex environments or the potential for widespread contamination. Finally, SADA version 4.1, along with a draft user's guide, was released in May 2005. This product is a tool for designing sampling programs to efficiently determine the extent of potential contamination to develop realistic survey plans.

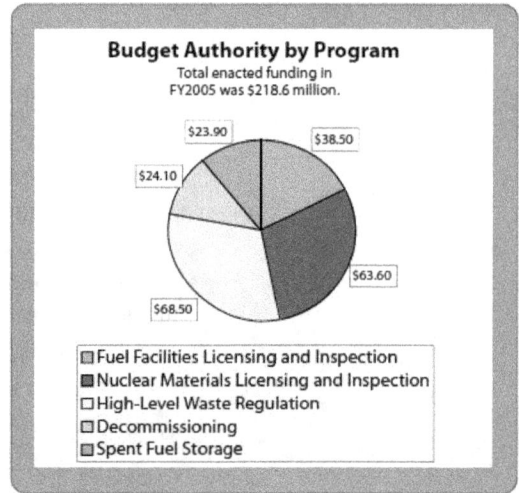

Budget Authority by Program
Total enacted funding in FY2005 was $218.6 million.

$23.90
$38.50
$24.10
$63.60
$68.50

- Fuel Facilities Licensing and Inspection
- Nuclear Materials Licensing and Inspection
- High-Level Waste Regulation
- Decommissioning
- Spent Fuel Storage

Figure 16

Funding for Achieving Goals

The Nuclear Materials and Waste Safety budget was $218.6 million in FY 2005. This went to the key activities of Fuel Facilities Licensing and Inspection, Nuclear Materials Users Licensing and Inspection, High-Level Waste Repository, Decommissioning and Low-Level Waste, and Spent Fuel Storage and Transportation (see Figure 16).

Program Evaluation

The NRC conducted an integrated materials performance evaluation of the Region I materials program in FY 2005. The integrated materials performance evaluation program is an ongoing oversight program designed to evaluate the quality, adequacy, and consistency of the NRC and Agreement State materials programs using a set of common performance indicators. The evaluation was conducted by a multi-disciplinary team of NRC and Agreement State personnel. The team found that the Region I operations are fully satisfactory with respect to the technical quality of licensing and inspections, the status of the inspection program, responses to incidents and allegations, and technical staffing and training. The Management Review Board supported the team's proposed findings and determined that the program is operating in a manner that adequately protects the public health and safety.

During FY 2005, an NRC team was established to perform an evaluation of the NRC's High-Level Waste Repository program. This evaluation examined the agencywide regulatory program for the proposed high-level waste repository at Yucca Mountain, Nevada, from the perspective of the Office of Management and Budget's Program Assessment Rating Tool (PART). This evaluation concluded that the program is well positioned to accomplish its objectives. The evaluation contains several recommendations for FY 2006 actions that will help the program prepare for a PART review, currently scheduled to occur in FY 2007.

Program Assessment Rating Tool

Over the past several years, the Office of Management and Budget has conducted reviews utilizing the Program Assessment Rating Tool of the Nuclear Materials and Waste Safety program activities. The following sections include a description of the activities, the Office of Management and Budget recommendations, and NRC's response and their impact upon program performance.

Spent Fuel Storage and Transportation Licensing and Inspection

In FY 2005, the NRC evaluated its Spent Fuel Storage and Transportation Licensing and Inspection activity using the Program Assessment Rating Tool promulgated by the Office of Management and Budget. The Office of Management and Budget rated the activity as "effective," which is the highest rating, and gave the activity an overall score of 89.

Fuel Facilities

Office of Management and Budget Recommendations

The Office of Management and Budget rated this activity as effective with an overall score of 89 in FY 2003 (Budget Year 2005). The activity earned high scores for Program Purpose and Design and for Program Management. The Office of Management and Budget noted that the purpose of the activity was clear, well-designed, and results-oriented. Also noted was that this activity has met all of its performance measures since the Government Performance and Results Act program reporting began in 1997. The Office of Management and Budget's recommendations included better linkage of budget requests to accomplishing annual and long-term goals and complete evaluation of performance measures in the organization's operating plan and revise them as necessary to support the safety performance measures in the NRC's FY 2004–FY 2009 Strategic Plan. Another recommendation is for more transparency in how allocation decisions are made and how the activity contributes to achievement of the agency's long-term goals, and complete the NRC's review of operating plan format and content to improve their effectiveness as management tools.

NRC Response

The NRC has developed better linkage of budget requests to its annual and long term goals through its initiative to define program outcomes and outputs that align with performance measures. Additionally, the NRC is working to improve its cost management capabilities to better align its costs with outcomes. The NRC also demonstrated via direct linkage of FY 2005 Operations Plan performance measures to the NRC FY 2004–FY 2009 Strategic Plan strategies for meeting the Strategic Plan objective and goals. Each of the operating plan's safety performance measures reference one or more of the Strategic Plan strategies for the agency's safety goal.

In addition, the NRC moved to implement costing to the NRC's two primary goals in the FY 2004–FY 2009 Strategic Plan (safety and security) beginning with the FY 2006 Performance Budget. In addition, the NRC has demonstrated better linkage of budget requests to agency goals through utilization of the common prioritization process for establishing the linkage between operational activities, including the resources allocated to support these activities, and the agency's strategic and long-term goals. The NRC's Fuel Cycle Licensing and Inspection program managers have responded to the Office of Management and Budget recommendation by linking operational activities and the agency's strategic and long-term goals in the revised operating plans.

The NRC also responded to completing a review of the organization's operating plan format and content. The scope of the project was separated into two phases to address: improvements that could be implemented in the short-term; and, improvements that would require longer-term planning and evaluation. The short-term improvement efforts were completed in December 2004 through the development of a performance reporting framework containing common reporting criteria and format. This format was implemented during the first quarter of FY 2005. The longer-term efforts to improve the efficiency of operating plans are currently being addressed by an agencywide workgroup.

Nuclear Materials Users Licensing and Inspection

Office of Management and Budget Recommendations

This Program Assessment Rating Tool review was conducted in FY 2004 for the FY 2006 Performance Budget. The Office of Management and Budget rated this activity as effective with an overall score of 93. The Office of Management and Budget recommendations include having the NRC provide a clearer demonstration of the contributions of specific activities to agency goals in the FY 2007 Performance Budget; create program goals that will support the mission of the agency; and schedule an evaluation of the program consistent with guidance in the Office of Management and Budget Circular A-11 prior to the submission of the FY 2007 Performance Budget. With respect to the third Office of Management and Budget recommendation, the NRC's Office of the Inspector General is currently conducting a review of the Nuclear Materials Users program area. The Office of the Inspector General report is expected in late FY 2005.

NRC Response

The NRC is currently developing milestones to address the Office of Management and Budget recommendations and the results will be included in next year's *Performance and Accountability Report.*

ADDRESSING THE PRESIDENT'S MANAGEMENT AGENDA

Overview

The President's Management Agenda prescribes Governmentwide initiatives to reform the U.S. Government to be more citizen-centered, results-oriented, and market-based, and to actively promote competition rather than stifling innovation. To achieve this goal, the Administration has identified five initiatives to improve Government performance in the areas of (1) strategic management of human capital, (2) budget and performance integration, (3) competitive sourcing, (4) expanded electronic Government, and (5) improved financial management. The NRC has responded to these Governmentwide initiatives in the following five sections, and discusses agency accomplishments during FY 2005 in each of the five areas, respectively.

Initiative 1: Strategic Management of Human Capital

Strategic Alignment

In FY 2005, the NRC continued the work begun in FY 2004 in its updated Strategic Human Capital and Workforce Restructuring Plan, which describes objectives and strategies for addressing the agency's human capital challenges. This plan aligns with the agency's FY 2004–FY 2009 Strategic Plan and with the agency's action plans for recruitment, training and development, and diversity management. In accordance with the plan, the NRC continues to identify future human capital investments through the agency's planning, budgeting, and performance management process.

Workforce Planning and Deployment

Various offices within the NRC improved operations when the agency completed changes in organizational structure. These changes included the realignment of functions, reductions in the span of control, elimination of unnecessary layers of management, and reorganizations. One initiative created a new low-level waste section to manage effectively the new duties stemming from the National Defense Authorization Act and the NRC's new responsibilities to monitor disposal actions and to consult with the Department of Energy on waste incidental to reprocessing determinations.

These improvements in organizational structure are integrated with continuing efforts to use the agency's strategic workforce planning process to improve workforce deployment, maintain technical capacity, and make informed decisions on human capital strategies for recruitment, development, and retention.

In addition, over the past four years, the NRC made significant improvements in the agency's strategic workforce planning methodology and system based on emerging needs and end user feedback. The Office of Personnel Management continues to cite the NRC's strategic workforce planning process and related Web-based application as an exemplary model for other Federal agencies.

Talent

Through partnerships between the program offices and the Office of Human Resources, the NRC employs human capital strategies to maintain the technical excellence of the NRC workforce, prepare for emerging work, address identified critical skill gaps, and meet and exceed the agency's human capital goals. These strategies include recruitment, relocation and retention incentives, student loan repayments, waivers of dual compensation limitations, partnerships with colleges and universities, the Cooperative Education Program, the Honor Law Graduate Program, the Graduate Fellowship Program, the Summer Employment Program, the Nuclear Safety Professional Development Program, rotational assignments, succession planning, mentoring, and training and development opportunities. These strategies have had a positive impact on the agency's efforts to recruit and retain staff with critical skills.

Leadership and Knowledge Management

The NRC uses succession planning, training and development, and knowledge management strategies to close identified critical skill gaps and to ensure continuity of leadership. The NRC continues to offer leadership competency development programs such as executive leadership seminars, the Senior Executive Service Candidate Development Program, leadership training for new supervisors and team leaders, and the Leadership Potential Program. These programs comprise a critical aspect of the NRC's succession and leadership development strategies by ensuring that leaders are prepared to assume entry-level, mid-level and senior-level leadership positions throughout the agency.

The NRC provides a wide variety of in-house, contracted, and online technical and professional training in the areas of reactor technology, engineering support, health physics, regulatory skills, communications, acquisition, and computer support. The NRC develops and conducts courses based on results from an annual training needs survey.

Performance Culture

Last year, the NRC implemented a new Senior Executive Service performance management system to improve its value as a management tool and to incorporate legislative changes as well as regulatory changes implemented by the Office of Personnel Management. The new system aligns individual executive performance expectations with the agency's Strategic Plan, Performance Budget, and office operating plans. The Office of Personnel Management and the Office of Management and Budget certified the NRC's Senior Executive System for FY 2004 and FY 2005, thus signifying that the NRC's system makes meaningful distinctions between the performances of various executives.

Accountability

The NRC continues to evaluate how well the agency is succeeding in achieving the human capital goals and outcomes in the areas of recruitment, staffing, retention, and training and development. In addition, the NRC staff briefs the Commission annually on the agency's human capital efforts.

Twice each year, the NRC analyzes and reports to the Commission on the status of workforce statistics by demographic groups over a five-year period. The analysis includes workforce size and composition, hires, attrition, rotational assignments, performance appraisals, and awards. These statistics are shared throughout the agency.

Initiative 2: Budget and Performance Integration

The NRC continues to make progress in achieving budget and performance integration in accordance with the President's Management Agenda. This progress includes adopting new outcome-based performance measures aligned with the agency's FY 2004–FY 2009 Strategic Plan, accurately monitoring program performance, and integrating performance information with associated costs. To address these initiatives, the NRC has pursued and completed a number of actions in FY 2005, as discussed in the following sections.

Integrating Planning and Budgeting

The NRC's planning, budgeting, and performance management process links the NRC's various budget accounts to the agency's primary goals of safety and security and clearly identifies the budgetary resources devoted to them. The agency's FY 2006 budget request identifies the alignment of resources to these two primary goals.

Full Cost Budget

NRC program managers currently receive cost reports that show the full cost of major programs. These reports allow managers to plan and manage their programs better throughout the budget year. The NRC's Performance Budget presents the "full cost" budget to achieve the agency's goals. The agency's FY 2005 budget request is the first budget submission in which the NRC has shown the full cost at the program level. The NRC will continue to refine the integration of outputs, goals, and assignment of full cost across programs as outlined in the Office of Management and Budget guidance for the FY 2006 budget.

Program Effectiveness

The NRC's Reactor Licensing activity and Spent Fuel Storage and Transportation Licensing and Inspection activities were evaluated using the Program Assessment Rating Tool promulgated by the Office of Management and Budget. The Spent Fuel Storage and Transportation Licensing and Inspection activity was rated effective, which is the Office of Management and Budget's highest rating. The Reactor Licensing activity was rated as moderately effective, the second highest rating. This finding resulted from the assessment that this activity needed more challenging annual measures and better efficiency measures. The NRC's experience from both reviews has yielded valuable insights for improving the measurement of the efficiency and effectiveness of its activities.

The NRC has modified its performance appraisal system for senior executives. The degree to which each senior executive's individual performance contributes to achieving the organization's goals and objectives is now an important part of their appraisal. The new system has been certified by both the Office of Management and Budget and the Office of Personnel Management as showing accountability for performance.

Initiative 3: Competitive Sourcing

One of the NRC's corporate management strategies is to acquire goods and services in an efficient manner. To achieve that, the NRC established output measures associated with the implementation of the competitive sourcing initiative under the President's Management Agenda, adopted a performance-based approach to contracting, and posted procurement synopses on the agency's Web site.

The NRC submitted its FY 2004 Federal Activities Inventory Reform Act inventory to the Office of Management and Budget in June 2004, and received approval from the Office of Management and Budget on November 16, 2004. That inventory identifies 248 commercial activity full-time equivalent units, which are available for public-private competition. The NRC published the inventory to its external Web site on November 17, 2004. One challenge to the 2004 commercial inventory was received. The NRC rendered its initial decision denying the challenge on February 10, 2005. The NRC denied the appeal to this decision on March 11, 2005. The NRC submitted its 2005 Federal Activities Inventory Reform Act inventory to the Office of Management and Budget on June 30, 2005.

The NRC conducted four business case analyses covering 18 full-time equivalents during FY 2004 to determine whether the selected commercial activities were appropriate for public-private competition based on the factors outlined in the NRC's Competitive Sourcing Plan. Based upon the Source Selection Authority's completed review of the four business case analyses, the NRC determined that it was not cost effective and, therefore, not appropriate to initiate public-private competitions for these activities. Three business case analyses are underway in FY 2005 and are planned to be completed by September 30, 2005, in accordance with NRC's Competitive Sourcing Plan.

The NRC continues to implement performance-based contracting for facility management services, data entry, information technology, and other support services. To give vendors a better understanding of contract requirements, the NRC includes such criteria as measurable performance requirements, quality standards, quality surveillance plans, and provisions for reducing the fee or price when the vendor fails to perform services as required. The NRC continues to exceed its target for expending eligible service contacting dollars through performance-based contracting. As a result, the agency has experienced improved vendor performance and lowered acquisition costs.

The NRC continues to post on its external Web site all required synopses and solicitations for acquisitions valued at more than $25,000.

Initiative 4: Expanded Electronic Government

The NRC continued to integrate and align its information technology investments with the Federal Government's Electronic Government program. The NRC uses Electronic Government services for payroll, security clearance, acquisition support, Governmentwide customer service, recruitment and training, and the NRC is currently implementing support for travel. In addition, for the 15 Presidential Priority initiatives that the NRC participates in through internal agency coordination, the NRC ensures alignment and consistency with Governmentwide standards and solution approaches. The NRC established procedures to avoid information technology investments that would duplicate other Federal Electronic Government programs and to take advantage of the SMARTBUY program. The NRC is participating in the Finance and Human Capital Lines of Business, and the agency is well positioned to take advantage of these programs because the NRC currently receives payroll and human resource services from Department of the Interior. The NRC is also participating in the Information Technology Security Lines of Business. The agency completed analysis of our Electronic Government implementation and alignment efforts as requested by the Office of Management and Budget and established key milestone dates, as appropriate. The NRC's Licensing Support Network system has been singled out by the Office of Management and Budget, and included in its annual Electronic Government report to Congress, as an example of a highly effective cross-agency initiative.

Enterprise Architecture

The NRC continued to make progress in embracing enterprise architecture. An enterprise architecture team was formed to ensure the timely coordination and completion of business-driven plans aligned with the Federal enterprise architecture for both the short and long term. The NRC is implementing business outreach activities through an Enterprise Architecture Communication Plan. The NRC is populating an automated enterprise architecture tool to capture and document the agency's enterprise architecture and to identify patterns and aid decisionmaking for information technology investments.

The NRC emphasizes enterprise architecture in its information technology systems development life cycle and is completing an integrated policy and process called the Project Management Methodology. The NRC project manager will have a single guide to meet both internal and external requirements. The Project Management Methodology consolidates existing NRC management directives and supporting processes for enterprise architecture, capital planning and investment control, systems development life cycle management methodology, and infrastructure development process model into one directive and handbook with an associated Web site, automated tool, and established processes. Besides fully integrating the consolidated policies and processes, the Project Management Methodology guide includes checkpoints for associated processes such as information technology security and records management. The Information Technology Business Council and Information Technology Advisory Council, comprising senior business managers, continue to play an important role in linking information technology investment decisions to the agency's mission and goals. The recently completed Enterprise Architecture Readiness Assessment provides useful information that serves our business strategic planning for information technology and enterprise architecture implementation efforts. The continuing accomplishments in enterprise architecture enable the building of better NRC business models that will provide the understanding necessary for us to more effectively solve business problems and provide better, more efficient information technology services.

Federal Information Security Management Act

The NRC's compliance with the requirements of the 2004 Federal Information Security Management Act was ranked third among all Government agencies and resulted in a grade of "B" issued by the House Committee on Government Reform's Subcommittee on Technology, Information Policy, Intergovernmental Relations, and the Census. In FY 2005, the NRC has increased efforts to conduct more rigorous independent review, testing, and evaluation of major system security plans. These increased efforts reveal previously undiscovered and unidentified security risks. In response, the NRC extended some system certification schedules to ensure full and complete system certification.

The NRC has an effective information technology security training and awareness program. All employees are required to complete an online information technology security training course, and NRC information systems security officers and other employees and support contractors with significant security responsibilities are required to complete a more advanced online technical security course. The NRC maintains an information technology security Web site and provides information to agency employees for the timely awareness of information technology security issues. The NRC has a robust incident reporting program in place and files monthly reports to the Federal Computer Incident Response Center.

E-Authentication Guidance

The Office of Management and Budget issued "E-Authentication Guidance for Federal Agencies," which updated earlier guidance under the Government Paperwork Elimination Act to ensure that online Government services are secure and protect privacy. This updated guidance directed agencies to conduct electronic authentication risk assessments and categorize all existing transactions and systems that require user authentication into four "identity assurance levels" by September 15, 2005. The NRC awarded a contract to complete these assessments for all electronic transactions in accordance with guidance promulgated by the National Institute of Standards and Technology. The NRC received an extension from OMB and will complete this effort by the end of December 2005.

Electronic Information Exchange—Minimizing the Burden on Business

The NRC maintains an electronic information exchange program, which provides for the transmission of digitally signed electronic documents to the NRC over the Internet. Information received in this manner can then be electronically disseminated directly into the agency's information systems. The NRC's Electronic Information Exchange program plays a major role in enabling the agency to meet the Government Paperwork Elimination Act requirement to allow the public the option of transacting business electronically with the agency. The NRC implemented system changes to accommodate the High-Level Waste activities. During FY 2005, approximately 30 legal briefs have been filed via Electronic Information Exchange in the High-Level Waste Pre-License Application Presiding Officer proceeding.

High-Level Waste Meta-System

Over the last three years, the NRC has been integrating several major agency applications—Agency Documents Access and Management System, Electronic Information Exchange, Electronic Hearing Docket, Digital Data Management System, and Licensing Support Network—and business processes to support licensing of the Department of Energy's nuclear waste disposal repository at Yucca Mountain, Nevada. In order to meet the challenges of licensing Yucca Mountain, the NRC has implemented new information systems and leveraged much of the existing information technology and information management architecture by enhancing computer applications, upgrading computing infrastructure, and improving business processes to provide a more robust, secure, and integrated environment. This collection of business processes, computer applications, and information technology infrastructure components is referred to as the High-Level Waste Meta-System. The High-Level Waste Meta-System's capability to support the High-Level Waste business process has been validated by performing iterative exercises of the entire business process. On June 2, 2005, the NRC conducted an Operational Readiness Review that resulted in the acceptance of Release 1 of the High-Level Waste Meta-System to support the High-Level Waste activities and adjudicatory proceedings.

Improvements to the NRC's Internal and Public Web Sites

The NRC participated in the American Customer Satisfaction Index and deployed the American Customer Satisfaction Index survey on our public Web site. The NRC will evaluate the statistics compiled from the survey results and measure how our public Web site performs in relation to other Government and private industry participants. The results will be used to identify areas that may need improvement.

The NRC launched a new public meeting notice system, accessible through our public Web site, which allows the public to search public meetings by docket number, facility name, meeting location, participants, and meeting dates. Agency stakeholders now can more easily identify and plan for meetings that are of interest to them.

The NRC has improved its emergency readiness to use its public Web site effectively. Staff from our emergency response, public affairs, and Web content management organizations collaborated to prepare appropriate procedures, Web page templates, and content that will be used during an emergency. The new procedures were tested and improved during two exercises in March and May 2005. In addition, the NRC began using a Web hosting service that prevents an overload in the event of a "denial of service" attack on our Web site or in emergency in which many members of the public try to access the agency Web site.

Sensitive Information Screening

Early this year, the NRC removed numerous documents from its publicly available records library (accessible from our public Web site) and screened these documents for information that could reasonably be expected to be useful to terrorists. The majority of these documents, with the exception of documents related to materials licenses, have been returned to public access after an extensive staff review effort and significant work by our IT staff to selectively remove and then restore segments of the information as the screening was completed.

Productivity Improvements

The NRC conducted a pilot to assess the viability of the Citrix Web Interface remote access system for high-speed remote access. As a result of the pilot, the Office of Information Services will offer a new service to the agency's remote workforce. Broadband Remote Desktop is a method where users can access the NRC network via their own existing high-speed broadband connection with their Internet service provider. The Broadband Remote Desktop will provide connectivity to the agency network via the Citrix servers at NRC headquarters or the Regional offices. Using a Web browser and the

NRC digital certificate, remote users can access their network files, NRC e-mail, and other network applications. The positive results of the pilot indicated that this technology would improve productivity of staff working offsite, and production implementation is currently underway.

The NRC upgraded the agency's firewall environment. The primary function of an agency firewall is to protect the NRC's network resources from Internet-based threats. This firewall is vendor-supported technology with an improved level of protection, availability, and performance for the NRC's Local Area Network/Wide Area Network. The firewall implementation resulted in better security, performance, and enforced NRC Internet policies.

The NRC developed a half-day hands-on computer course "Making 508-Compliant Government Purchase Card Decisions" in the agency's Professional Development Center. The course is intended for NRC staff who need to use Government purchase cards to purchase computers, software, telecom, faxes, calculators, and videos that comply with Section 508 of the Americans with Disabilities Act. This course provides users with a systematic process tool for identifying which items meet the requirements. The use of these tips and tools will result in a productivity improvement for Government credit card holders.

Initiative 5: Improved Financial Management

Financial Management Systems

The NRC's financial systems strategy is to improve business processes, systems performance, and access to information while reducing life-cycle costs by relying on commercially available software and cross-service providers wherever possible. The NRC's core accounting, payroll, and human resources systems are cross-serviced by a Federal agency Center of Excellence. The remaining internally maintained and managed financial systems are periodically reviewed to identify ways to improve performance, interface with other systems, and utilize cross-servicing, as appropriate. The agency also provides electronic access to daily financial transaction data and reports, as well as agency standard cost ratio and performance data. Our current systems satisfy operational and reporting requirements and provide timely, accurate, and useful information to agency managers.

The NRC's financial systems are in substantial compliance with the Improvement Act, except for its Fee Billing System and the payroll and core accounting systems cross-serviced by the Department of Interior (DOI) National Business Center (NBC). These systems are in substantial noncompliance with Federal financial management system requirements.

Improvements were also made in the cost accounting system in FY 2005. An obligation model was created that will allow tracking costs by obligation which resource managers use to make decisions regarding resource utilization. New reports were created in the Cost Accounting System, which is used to monitor charges to the Nuclear Waste Fund by program offices using a full cost methodology. Also, the Cost Accounting System was updated to reflect the FY 2005 budget structure. All financial and managerial cost reports were issued on time or ahead of schedule for FY 2005.

Accurate and Timely Financial Information

The NRC received an unqualified opinion on the FY 2005 financial statements, and the FY 2004 *Performance and Accountability Report* earned the agency a Certificate of Excellence in Accountability Reporting from the Association of Government Accountants.

Integrated Financial and Performance Management Systems

The NRC has achieved a high level of financial systems integration, which supports the agency's day-to-day operations. To achieve this integration, core accounting is interfaced with the cost accounting, payroll, and fee billing systems. The agency also provides electronic access to daily financial transaction data and periodic summary reports for management use. Senior managers receive monthly budget execution reports as well as agency standard cost ratio and performance data.

Annual Financial Statements and Internal Controls

The NRC earned an unqualified audit opinion on the agency's financial statements in FY 2005. The NRC will continue to pursue actions that will result in the issuance of financial statements with unqualified audit opinions and no material internal control weaknesses. During FY 2005, NRC continued efforts to eliminate the auditor-identified material internal control weakness related to the Fee Billing System. NRC implemented improvements to the fee billing process and resolved two reportable conditions, but further corrective action is needed to address the remaining three.

In order to promote a high level of data integrity, the NRC has a robust system of internal controls designed to ensure that financial data are entered in a timely and accurate manner. The system of internal controls requires monthly reconciliation of data and quarterly certification by managers throughout the agency. The agency has an established program for routinely assessing performance and financial information. Annually, managers are required to provide reasonable assurance that effective controls are in place to ensure the integrity of their program and financial operations. These reasonable assurance assessments are reviewed by an executive agency management group, which in turn provides assurance to the Chairman of the Commission. This is the basis for the Chairman's assurance statement contained in the agency's annual Performance and Accountability Report.

DATA SOURCES AND QUALITY

The NRC's data collection and analysis methods are driven largely by the regulatory mandate that Congress entrusted to the agency. Specifically, the NRC's mission is to regulate the Nation's civilian use of byproduct, source, and special nuclear materials to ensure adequate protection of public health and safety, protect the environment, and promote the common defense and security. In undertaking this mission, the NRC oversees nuclear power plants, nonpower reactors, nuclear fuel facilities, interim spent fuel storage, radioactive material transportation, disposal of nuclear waste, and the industrial and medical uses of nuclear materials. Section 208 of the Energy Reorganization Act of 1974, as amended, requires the NRC to inform Congress of incidents or events that the Commission determines to be significant from the standpoint of public health and safety. The NRC developed the abnormal occurrence criteria to comply with the legislative intent of the Act to determine which events should be considered "significant." Based on those criteria, the NRC prepares an annual "Report to Congress on Abnormal Occurrences" (NUREG-0090, Vol. 26), which is available on the agency's public Web site at http://www.nrc.gov/reading-rm/doc-collections/nuregs/staff/sr0090.

One important characteristic of this report is that the data presented normally originate from external sources such as Agreement States and NRC licensees. The NRC believes that these data are credible because (1) agency regulations require Agreement States, licensees, and other external sources to report the necessary information; (2) the NRC maintains an aggressive inspection program that, among other activities, includes auditing licensee programs and evaluating Agreement State programs to ensure that they are reporting the necessary information as required by the agency's regulations; and (3) the agency has established procedures for inspecting and evaluating licensees.

The NRC employs multiple database systems to support this process, including the Licensee Event Report Search System, the Accident Sequence Precursor Database, the Nuclear Materials Events Database, and the Radiation Exposure Information Report System. In addition, nonsensitive reports submitted by Agreement States and NRC licensees are available to the public through the NRC's Agencywide Documents Access and Management System, which is accessible through the agency's public Web site at http://www.nrc.gov.

The NRC has established procedures for the systematic review and evaluation of events reported by NRC and Agreement State licensees. NRC's objective is to identify events that are significant from the standpoint of public health and safety based on criteria that include specific thresholds. The NRC verifies the reliability and technical accuracy of event information reported to the agency. The NRC periodically inspects licensees and reviews Agreement State programs. In addition, NRC headquarters, the Regional offices, and Agreement States hold periodic conference calls to discuss event information. Events identified as meeting the abnormal occurrence criteria are validated and verified by all applicable NRC headquarters program offices, Regional offices, and agency management before being reported to Congress.

Data Security

Data security is ensured by the agency's automated information security program, which provides administrative, technical, and physical security measures to protect the agency's information, automated information systems, and information technology infrastructure. Specifically, these measures include the policies, processes, and technical mechanisms used to protect classified information, unclassified safeguards information, and sensitive unclassified information that are processed, stored, or produced on the agency's automated information systems. Data security for information maintained outside the NRC's infrastructure is provided by the hosting contractor or organization.

For major systems, the NRC ensures compliance with agency standards through independent reviews conducted under the Federal Information Security Management Act. The NRC's Office of the Inspector General completed its independent assessment of the agency's implementation of the Act on September 30, 2004. Through that assessment, the Office of the Inspector General found that the NRC has continued to improve its security program by completing a majority of program and system level corrective actions identified in the FY 2003 Federal Information Security Management Act review, including additional corrective actions identified through FY 2004, and developing processes and procedures for updating the NRC system inventory and implementing security configurations on NRC servers.

Performance Data Completeness and Reliability

In order to manage for results, it is essential for the NRC to assess the completeness and reliability of our performance data. Comparisons of actual performance with the projected levels are possible only if the data used to measure performance are complete and reliable. Consequently, the Reports Consolidation Act of 2000 requires the Chairman of the NRC to assess the completeness and reliability of the performance data used in this report. In addition, the Office of Management and Budget Circular A-11 specifically describes how Federal agencies should assess the completeness and reliability of their performance data.

Data Completeness

The Office of Management and Budget considers data to be complete if an agency reports actual performance data for every performance goal and indicator in the annual plan. Actual performance data may include preliminary data if those are the only data available when the agency sends its report to the President and Congress. The data presented in this report meet these requirements for data completeness, in that we have reported actual or preliminary data for every strategic and performance goal measure.

The actual data for strategic and performance goal measures covers the entire fiscal year for 2005 unless otherwise noted in the *Performance and Accountability Report*.

Data Reliability

The Office of Management and Budget considers data to be reliable when agency managers and decisionmakers do not demonstrate either a refusal or a marked reluctance to use the data in carrying out their responsibilities. The data presented in this report meet this requirement for data reliability in that the NRC's managers and decisionmakers regularly use the reported data on an ongoing basis in the course of their duties.

Improvements in Performance Data

The NRC analyzed the data verification procedures for the agency's performance measures during FY 2005. This analysis consisted of an evaluation of all data collection, analysis, and reporting procedures for completeness, accuracy, consistency, and timeliness. The analysis also included an evaluation of NRC management controls, which ensure that the reported data are valid and reliable. As a result, the NRC believes that its performance data are both valid and reliable.

A more complete discussion concerning the validation and verification of the NRC's performance measures is provided in the agency's Performance Budget for FY 2006 (NUREG-1100, Vol. 21), which the Commission submitted to Congress in February 2005. The Performance Budget is available on the NRC's public Web site at http://www.nrc.gov/reading-rm/doc-collections/nuregs/staff/sr1100/. Appendix IV to the NRC's Performance Plan provides an extensive explanation of the NRC's data verification and validation procedures for each performance measure.

The NRC makes performance data accessible to citizens through the public Web site. For example, a citizen who wanted to verify or know more about licensee event reports, which provide the raw data for most of our performance measures, could simply retrieve any or all of those reports through the NRC's Agencywide Documents Access and Management System (ADAMS), accessible through our public Web site at http://www.nrc.gov/reading-rm/adams.html, by searching for "licensee event report."

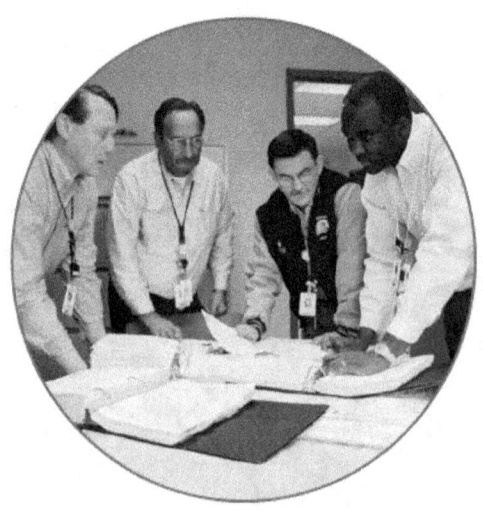

. . . committed
to effective
and efficient
management of
our resources . . .

A MESSAGE FROM THE CHIEF FINANCIAL OFFICER

I am pleased to present the U.S. Nuclear Regulatory Commission's financial statements for FY 2005 as an integral part of the agency's *FY 2005 Performance and Accountability Report*. Our independent auditor has rendered an unqualified opinion on our financial statements, attesting to the fact that NRC's financial statements are fairly presented and demonstrate discipline and accountability in the execution of our responsibilities as stewards of the American taxpayers' dollars.

As of September 30, 2005, the financial condition of the NRC is sound with respect to having sufficient funds to meet its mission and having adequate control of these funds to ensure our budget authority is not exceeded. We successfully collected approximately 99 percent of the agency's budget that is subject to fee recovery from NRC licensees and our year-end delinquent debt was minimal, less than our goal of one-half of one percent of the fees collected. Ninety-six percent of payments to commercial vendors, subject to the Prompt Payment Act, were made on-time with less than one-half of one percent made erroneously. We have also received excellent ratings for maintaining our fund balance with Treasury and for our timely and accurate reporting.

During FY 2005, we continued our efforts to eliminate the auditor-identified material internal control weakness related to the Fee Billing System. We implemented improvements to the fee billing process but further corrective action is needed (Chapter 1, *Audit Results*). In addition, we have resolved two reportable conditions and are working to address the remaining three. We identified three systems, Fee Billing System and two systems provided under an e-Government cross-servicing arrangement (core accounting and payroll), to be in substantial noncompliance with Federal financial management system requirements. We developed a remediation plan for the Fee Billing System targeted to achieve compliance in 2007 and are monitoring the corrective actions being taken by the cross-service provider to bring their systems into compliance.

The NRC is committed to effective and efficient management of its resources. Our goals and strategies for improving financial management are centered on maintaining unqualified audit opinions, eliminating our material internal control weakness, upgrading financial systems to meet Federal requirements, meeting new reporting requirements, and implementing e-Government initiatives. We continue to be successful in ensuring that our operations provide timely and reliable information that is used to promote results, accountability, and efficiency. This has been possible through the efforts and teamwork of program, financial management, and audit staff.

I anticipate another productive year in FY 2006 and a continuation of the same high level of financial services that resulted in our past successes. While we make progress, we are mindful of our support role in getting an unqualified audit opinion on the *Financial Report of the United States Government*.

Jesse L. Funches

November 15, 2005

AUDITORS' REPORTS

**UNITED STATES
NUCLEAR REGULATORY COMMISSION**
WASHINGTON, D.C. 20555-0001

OFFICE OF THE
INSPECTOR GENERAL

November 10, 2005

MEMORANDUM TO: Chairman Diaz

FROM: Hubert T. Bell
                                    Inspector General

SUBJECT: RESULTS OF THE AUDIT OF THE
                                    UNITED STATES NUCLEAR REGULATORY COMMISSION'S
                                    FINANCIAL STATEMENTS FOR FISCAL YEARS 2005
                                    AND 2004 (OIG-06-A-01)

The Chief Financial Officers Act of 1990, as amended, (CFO Act) requires the Inspector
General (IG) or an independent external auditor, as determined by the IG, to annually
audit the United States Nuclear Regulatory Commission's (NRC) financial statements
in accordance with applicable standards. In compliance with this requirement, this
memorandum transmits the following R. Navarro & Associates, Inc. Auditors' Reports:

- Independent Auditors' Report on the FYs 2005 and 2004 Financial Statements,

- Report on the Effectiveness of Internal Control over Financial Reporting, and

- Report on Compliance with Laws and Regulations.

Objective of a Financial Statement Audit

The objective of a financial statement audit is to determine whether the financial
statements are free of material misstatement. An audit includes examining, on a test
basis, evidence supporting the amounts and disclosures in the financial statements. An
audit also includes assessing the accounting principles used and significant estimates
made by management as well as evaluating the overall financial statement presentation.

R. Navarro & Associates' examination was made in accordance with generally accepted auditing standards, *Government Auditing Standards* issued by the Comptroller General of the United States, and Office of Management and Budget (OMB) Bulletin No. 01-02, *Audit Requirements for Federal Financial Statements*. The audit included obtaining an understanding of the internal controls over financial reporting and testing and evaluating the design and operating effectiveness of the internal controls. Because of inherent limitations in any internal control, there is a risk that errors or fraud may occur and not be detected. Also, projections of any evaluation of internal control over financial reporting to future periods are subject to the risk that the internal control may become inadequate because of changes in conditions, or that the degree of compliance with policies or procedures may deteriorate. The risk of fraud is inherent to many of NRC's programs and operations.

Results of Audit

The results are as follows:

Financial Statements

- FYs 2005 and 2004 — Unqualified opinion

FY 2005 Internal Controls

- Qualified opinion
- Reportable Conditions:
    - Fee Billing System (Also represents a **Continuing Material Weakness**)
    - Monitoring of Accounting for Internal Use Software (Continuing Condition)
    - Information System-Wide Security Controls (New Condition)
    - Financial Controls over Disbursements (New Condition)

FY 2005 Compliance with Laws and Regulations

- Noncompliances:
    - Part 170 Hourly Rates (Continuing Noncompliance)
    - Fee Billing System (Continuing **Substantial Noncompliance**)
    - Information System-Wide Security Controls (New **Substantial Noncompliance**)

OIG Evaluation of R. Navarro and Associates, Inc., Performance

To fulfill our responsibilities under the CFO Act and related legislation for ensuring the quality of the audit work performed, we monitored R. Navarro & Associates' audit of NRC's FYs 2005 and 2004 financial statements by:

- Reviewing their approach and planning of the audit,

- Evaluating the qualifications and independence of its auditors,

- Monitoring the progress of the audit at key points,

- Examining the workpapers related to planning and performing the audit and assessing NRC's internal control, Reviewing R. Navarro & Associates' audit reports to ensure compliance with *Government Auditing Standards* and OMB Bulletin No. 01-02,

- Coordinating the issuance of the audit reports, and

- Performing other procedures that we deemed necessary.

R. Navarro & Associates, Inc., is responsible for the attached auditors' reports, dated November 4, 2005, and the conclusions expressed therein. The Office of the Inspector General (OIG) is responsible for technical and administrative oversight regarding the firm's performance under the terms of the contract. Our review, as differentiated from an audit in conformance with *Government Auditing Standards*, was not intended to enable us to express, and accordingly we do not express an opinion on NRC's financial statements, the effectiveness of its internal control over financial reporting, or NRC's compliance with laws and regulations. However, our monitoring review, as described above, disclosed no instances where R. Navarro & Associates, Inc., did not comply with applicable auditing standards.

Performance Reporting

As required by OMB Bulletin No. 01-02, with respect to internal control related to performance measures determined by management to be key and reported in the *Management's Discussion and Analysis*, we obtained an understanding of the design of significant internal controls relating to the existence and completeness assertions. Our procedures were not designed to provide assurance on internal control over performance measures and, accordingly, we do not provide an opinion thereon.

Meeting with the Chief Financial Officer

At the exit conference on November 7, 2005, representatives of the Office of the Chief Financial Officer, OIG, and R. Navarro & Associates, Inc., discussed the issues in the report.

Comments of the Chief Financial Officer

In his response, the CFO generally agreed with the auditors' recommendations. We will follow-up on the CFO's planned corrective actions during FY 2006. The full text of the CFO's response follows this report.

We appreciate NRC staff's cooperation and continued interest in improving financial management within NRC.

Attachment: As stated

cc: Commissioner McGaffigan
        Commissioner Merrifield
        Commissioner Jaczko
        Commissioner Lyons

R. NAVARRO
& ASSOCIATES, INC.
CERTIFIED PUBLIC ACCOUNTANTS

2831 Camino Del Rio South, Suite 306
San Diego, California 92108
(619) 298-8193

Chairman Nils J. Diaz
U.S. Nuclear Regulatory Commission
Washington, DC

In our audits of the U.S. Nuclear Regulatory Commission (NRC), we found:

- The balance sheets of NRC as of September 30, 2005, and 2004, and the related statements of net cost, statements of changes in net position, statements of budgetary resources, and statements of financing for the fiscal years then ended are presented fairly, in all material respects, in conformity with accounting principles generally accepted in the United States of America;

- Except for the material weakness over the Fee Billing System, the effectiveness of internal control over financial reporting was fairly stated as of September 30, 2005, in compliance with the internal control objectives in the Office of Management and Budget (OMB) Bulletin No. 01-02, *Audit Requirements for Federal Financial Statements*. The Bulletin requires that transactions be properly recorded, processed, and summarized to permit the preparation of the financial statements in accordance with accounting principles generally accepted in the United States of America and that assets be safeguarded against loss from unauthorized acquisition, use or disposal; and

- The NRC continues to be noncompliant with the provisions of OMB Circular A-25, *User Charges*, for Part 170 fees. Additionally, NRC continues to have a substantial noncompliance related to the Fee Billing System, and, in the current year, we are reporting an Improvement Act substantial noncompliance with the security controls over applications operating at a cross-servicing agency.

The following sections outline each of these conclusions in more detail.

INDEPENDENT AUDITOR'S REPORT ON THE FINANCIAL STATEMENTS

We have audited the accompanying balance sheets of NRC as of September 30, 2005, and 2004, and the related statements of net cost, statements of changes in net position, statements of budgetary resources, and statements of financing for the fiscal years then ended. These financial statements are the responsibility of NRC's management. Our responsibility is to express an opinion on these financial statements based on our audits.

We conducted our audits in accordance with auditing standards generally accepted in the United States of America, the standards applicable to financial audits contained in *Government Auditing Standards*, issued by the Comptroller General of the United States, and OMB Bulletin No. 01-02. Those standards require that we plan and perform the audits to obtain reasonable assurance about whether the financial statements are free of material misstatement. An audit includes examining, on a test basis, evidence supporting the amounts and disclosures in the financial statements. An audit also includes assessing the accounting principles used and significant estimates made by management, as well as evaluating the overall financial statement presentation. We believe that our audits provide a reasonable basis for our opinion.

Matters of Emphasis

Classification of Costs

OMB Circular A-136, *Financial Reporting Requirements*, provides guidance to Federal agencies for presenting program costs classified by intragovernmental and public components. The basis for classification relies on the concept of who received the benefits of the costs incurred (i.e., private sector licensees versus Federal licensees) rather than who was paid. However, following the advice of OMB, NRC classified the costs on the Statement of Net Cost using an underlying concept of who was paid. Furthermore, OMB Circular A-136 requires that the Statement of Net Cost be presented using full program costs by output. The agency presents its costs aggregated by strategic plan programs.

U.S. Department of Energy Expenses

NRC's principal statements include reimbursable expenses of the U.S. Department of Energy (DOE) National Laboratories. For the years ended September 30, 2005, and 2004, NRC's Statements of Net Cost include approximately $68.7 and $77.2 million, respectively, of reimbursed expenses. Our audits included testing these expenses for compliance with laws and regulations applicable to NRC. The work placed with DOE is under the auspices of a Memorandum of Understanding between NRC and DOE. The examination of DOE National Laboratories for compliance with laws and regulations is DOE's responsibility. This responsibility was further clarified by a memorandum of the Government Accountability Office's (GAO) Assistant General Counsel, dated March 6, 1995, where he opined that "...DOE's inability to assure that its contractors' costs [National Laboratories] are legal and proper...does not compel a conclusion that NRC has failed to comply with laws and regulations." DOE also has the cognizant responsibility to assure audit resolution and should provide the results of its audits to NRC.

In our opinion, the financial statements referred to above and included in NRC's performance and accountability report present fairly, in all material respects, the financial position as of September 30, 2005, and 2004, and its net cost, changes in net position, budgetary resources, and reconciliations of net cost to budgetary resources for the fiscal years then ended in conformity with accounting principles generally accepted in the United States of America.

Report on the Effectiveness of Internal Control Over Financial Reporting

We have examined the effectiveness of NRC's internal control over financial reporting, as of September 30, 2005, based on the criteria in OMB Bulletin No. 01-02. The Bulletin requires management to establish internal accounting and administrative controls to provide reasonable assurance that transactions are properly recorded, processed, and summarized to permit the preparation of the financial statements in accordance with accounting principles generally accepted in the United States of America and that assets be safeguarded against loss from unauthorized acquisition, use or disposal. NRC's management is responsible for maintaining effective internal control over financial reporting. Our responsibility is to express an opinion on the effectiveness of internal control based on our examination.

Our examination was conducted in accordance with the attestation standards established by the American Institute of Certified Public Accountants (AICPA); the standards applicable to financial statement audits contained in *Government Auditing Standards*, issued by the Comptroller General of the United States; and OMB Bulletin No. 01-02. Accordingly, we obtained an understanding of the internal control over financial reporting, tested and evaluated the design and operating effectiveness of internal control, and performed such other procedures as we considered necessary in the circumstances. We believe that our examination provides a reasonable basis for our opinion.

Because of inherent limitations in any internal control, misstatements due to error or fraud may occur and not be detected. Also, projections of any evaluation of internal control over financial reporting to future periods are subject to the risk that the internal control may become inadequate because of changes in conditions, or that the degree of compliance with policies or procedures may deteriorate.

We identified continuing significant deficiencies in the Fee Billing System. The system in place does not meet the requirements of sound internal control over financial reporting as provided in OMB Bulletin No. 01-02, nor is the system's design compliant with the requirements of the Joint Financial Management Improvement Program (JFMIP—Effective December 2004, JFMIP ceased to become a standalone entity). Its functions are now under the auspices of OMB and the Chief Financial Officers' Council) for Revenue Systems. We believe such a condition represents a material weakness. A material weakness is a reportable condition that precludes the NRC's internal control from providing reasonable assurance that material misstatements in the financial statements will be prevented and detected on a timely basis.

In our opinion, except for the effect of the material weakness described in the preceding paragraph, NRC has maintained, in all material respects, effective internal control over financial reporting as of September 30, 2005, based on the internal control objectives listed in OMB Bulletin No. 01-02.

Additionally, we noted certain matters involving the internal control and its operation that we consider to be reportable conditions under standards established by the AICPA and OMB Bulletin No. 01-02. A reportable condition is a matter coming to our attention relating to significant deficiencies in the design or operation of the internal control that, in our judgment, could adversely affect the agency's ability to meet the internal control objectives described above. We identified four reportable conditions; NRC needs to

(1) improve the Fee Billing System, (2) improve monitoring of accounting for internal use software, (3) strengthen information system-wide security controls, and (4) strengthen financial controls over disbursements. The Fee Billing System condition is considered a material weakness.

A material weakness, as defined by the AICPA and OMB Bulletin No. 01-02, is a reportable condition in which the design or operation of one or more of the internal control components does not reduce to a relatively low level the risk that misstatements caused by error or fraud in amounts that would be material in relation to the principal financial statements being audited may occur and not be detected within a timely period by employees in the normal course of performing their assigned functions. We believe that the reportable conditions that follow, except for the Fee Billing System, are not material weaknesses as defined by the AICPA and OMB Bulletin No. 01-02.

Fee Billing System

During the assessment of internal controls in FY 2004, we identified significant deficiencies in the NRC's Fee Billing System, as described below. The agency has put forth a significant effort to address the issues reported in the prior year; however, greater emphasis and demonstrated sustainable business process improvements must be provided to remediate the material weakness.

The Omnibus Budget Reconciliation Act (OBRA-90), Public Law 101-508, as amended, requires that NRC recover, through fee billing, a percentage of its budget authority in each fiscal year, less amounts appropriated from the Nuclear Waste Fund. In FYs 2005 and 2004, the recovery percentage was 90 and 92 percent, respectively. In order to meet this requirement, the NRC assesses two types of fees to recover its budget authority. Annual fees are assessed under 10 CFR Part 171 for nuclear facilities and materials licensees, commonly known as Part 171 fees. Other fee types include license, inspection, and other services, established in 10 CFR Part 170 under the authority of the Independent Offices Appropriation Act (IOAA). The Part 170 fees are assessed to recover NRC's costs of providing individually identifiable services to specific applicants and licensees.

The conditions reported in the prior year resulted from several deficiencies, (1) inadequate acceptance testing of software modifications, (2) intensive manual processes, and (3) the lack of comprehensive quality assurance procedures over the billing process. In the current year, the agency has continued to identify underbilling problems with the Fee Billing System, indicating the need to more diligently document and design internal controls for each operating aspect of the system.

The GAO's *Standards for Internal Control in the Federal Government* state, "Internal control should generally be designed to assure that ongoing monitoring occurs in the course of normal operations. It is performed continually and is ingrained in the agency's supervisory activities, comparisons, reconciliations, and other actions people take in performing their duties."

The following examples provide insights into the agency's progress and current condition in addressing (1) intensive manual processes, (2) lack of comprehensive quality assurance procedures, and (3) fee billing feeder processes.

Intensive Manual Processes

Due to the age and design of the Fee Billing System, NRC has evolved over the years into an operating style characterized by over-reliance on a small team to prepare, review, and issue billings on a monthly and quarterly basis. The License Fee Team employs various manual processes to compensate for the lack of flexibility in the legacy fee billing system. The system does not have the ability to give the agency drill down capacity to review billing questions. In particular, the system does not provide automated audit trails from the initial source of the transaction (i.e., billable hours) to the development of an invoice.

Over the last 2 fiscal years, the agency performed an assessment of the Fee Billing System and concluded, "...that the existing nine systems that collectively comprise the Fee Systems will not fully support fee billing and will not promote consistency across the agency. Streamlining, automating, and improving its fee systems and processes with modern and integrated technology and processes will be critical to the agency, its staff, and its customer going forward." The agency prepared a remediation plan describing actions designed to replace the existing system. Deployment of the replacement solution is planned for FY 2008.

The lack of system functionality coupled with the age of the system and its reliance on manual intervention continues to result in an Improvement Act substantial noncompliance.

Lack of Comprehensive Quality Assurance Procedures

During the current year, the agency developed quality assurance procedures to reconcile the completeness of Part 170 (hourly) invoices to the license fee reports produced by the Fee Billing System. The reports provide the amounts available for billing. Late in the fiscal year, an accompanying reconciliation was performed using the new quality assurance procedures. However, the agency did not address several other reconciliation points that are essential to the internal control over fee billings.

For example, the quality assurance procedures did not address the completeness of billable contract costs as compared to contract payments made to vendors. The procedures also do not provide for a review of the reliability and completeness of data inputs from sources outside the Office of the Chief Financial Officer's (OCFO) business domain, which are integral to the reliability of invoices. Regional and technical offices such as Nuclear Materials Safety and Safeguards are the feeder source for license fee activities. This data is fundamental to the mapping of license fee rates in the billing preparation process. We commend the agency for its prompt action in developing some quality assurance procedures; however, much more needs to be done to mitigate known design and system risks of the legacy system and to assert to the completeness and reliability of the fee billing process.

Fee Billing Feeder Processes

In the current year, the agency identified several instances of underbilling, some of which date back to FY 1991. Although the net value of these unbilled accounts receivable does not have a material impact on the Balance Sheet, these instances highlight risks that are present in the Fee Billing System. The instances identified demonstrate the need to validate the feeder data from offices outside the CFO's business domain. The issues identified impact both Part 170 (hourly) and Part 171 (annual) fee billings and were identified during tasks related to data conversion, cost accounting data analysis, and policy research, as follows:

- **Data Conversion**—The agency is undergoing a conversion of the system used by the Office of Nuclear Materials Safety and Safeguards to track the licensee information by type of license. During this process, the universe of licensees was reviewed for completeness. The review identified several licensees that were inadvertently dropped from the universe of billable annual materials licensees. The information captured in this system by the program office is a key element of the data sources necessary for OCFO to identify licensees available for billing. The OCFO does not have procedures to verify the completeness of this data prior to initiating a billing cycle. The agency performed an analysis of the listing of dropped licensees and identified approximately $911 thousand of unbilled fees. The net effect on the Balance Sheet after providing for the allowance for doubtful accounts is approximately $197 thousand.

- **Cost Accounting Data Analysis**—During the third quarter billing cycle, the Cost Accounting Team identified transactions that were assigned to suspended activity codes. Subsequently, the license fee staff researched these transactions and found that billable time (Part 170–hourly fees) had not been billed. Program activity codes are used to capture agency costs within NRC's established budgetary program framework. The program activity codes were suspended because the general ledger tables containing viable billable codes had not been properly synchronized. At the start of each fiscal year, program activity codes are set up for all anticipated activities; however, as the year progresses new activity codes may be assigned, thereby triggering table maintenance issues. The unbilled amount was approximately $20 thousand. The agency has indicated it will develop a process to verify that all tables are properly synchronized prior to starting a billing cycle.

- **Policy Research**—During research of a policy question on billing of project managers' time (Part 170–hourly fees), the license fee staff found that billable time assigned by Nuclear Materials Safety and Safeguards managers was not included in billed amounts. Although the proper program activity codes were established at the beginning of the year, the contractor who supports the billing process was not advised of the code for general and administrative activities. As a result, $50 thousand was not billed.

Although not material to these financial statements, the unbilled amounts illustrate the need for improved quality assurance procedures over the billing preparation process. The agency has indicated it will pursue billing and collection of each amount described in these examples.

Recommendations

1. The CFO should direct an assessment of all aspects of the Fee Billing System to ensure that the remediation plan is updated as necessary and implemented in a timely manner to enhance the controls over fee billing processes.

2. The CFO should define, design, and implement compensating controls over the fee billing system, while the system is being considered for redesign.

3. As the CFO identifies needed improvement of internal controls outside OCFO's business domain, there should be coordination and collaboration with the Executive Director for Operations as to how, when and to what extent the internal controls should be strengthened in operational program feeder systems, relied upon by OCFO for billing preparation purposes.

Monitoring of Accounting for Internal Use Software

Although the OCFO has made strides with policy development and training, we continue to identify costs incurred and not recorded by OCFO for internal use software. OCFO's management control structure is designed to rely heavily on project managers to inform OCFO of time and costs expended in the software development phase. OCFO has not been fully successful in identifying projects through their monitoring procedures in order to ensure the completeness or reasonableness of the project manager's information.

Federal Accounting Standards Advisory Board issued Statements of Federal Financial Accounting Standards (SFFAS) No. 10, *Accounting for Internal Use Software,* effective October 1, 2000. SFFAS No. 10 defines three software life-cycle phases, planning, development, and operations. Paragraph 16 requires, "For internally developed software, capitalized cost should include the full cost (direct and indirect cost) incurred during the development phase." The Statement defines full cost to include salaries of programmers, project managers, administrative personnel, and associated employee benefits and outside consultants' fees.

Our review of the agency's practices for accounting for internal use software projects continues to identify the following inconsistencies:

- Contractor cost incurred on projects for work performed, but for which NRC has not been billed, were not being captured and capitalized;

- Project managers, in some instances, were not coding their time appropriately during the development phase of their projects; and

- Labor certifications, in some instances, were not being completed, signed and/or were being completed late.

For example, in the current year our audit procedures identified several projects where the costs were not properly recorded. At September 30, 2004, several projects were capitalized at $480 thousand; however, we noted that the actual cost of the projects was in excess of $700 thousand. The OCFO did not have a business process to collect invoices from contractors involved in the development process in order to more accurately capture project costs. The June 30, 2005, quarterly financial statements reflected the correction of these asset values.

Recommendation

4.   The CFO should continue to promote strengthening internal use software practices, in order to encourage project managers to comply with procedures in effect governing the completeness of new and existing development initiatives.

Information System-wide Security Controls

A recent report issued by the Office of Inspector General (OIG) (Report No. OIG-05-A-21) identified risks in the agency's information security environment. The report identifies various conditions placing the agency in an "at risk" position. The following is a partial list of the issues reported:

* A majority of the information systems (19 of 27) are under an interim authorization to operate and, therefore are not considered certified;

* Agency information system security self-assessments were not performed timely;

* Annual contingency planning is not being performed; and

* Oversight of other contractor systems is lacking.

The OIG's report states, "NRC's general support systems have not had a complete certification and accreditation performed in the past 3 years. Therefore, the agency does not know whether the security controls for these general support systems are adequate, creating unknown potential risk. As a result, all NRC information systems that depend on the security controls provided by these general support systems inherit that unknown potential risk."

The primary agency financial reporting systems include cost accounting, human resources management system, fees and two systems outsourced with Department of Interior's National Business Center (DOI-NBC). The two outsourced systems are the Federal Financial System (i.e., the general ledger application) and Federal Personnel and Payroll System (i.e., the payroll application). These systems operate or have access protocols on the NRC's general support system, which has been identified as vulnerable, since the general support system had a lapsed authorization to operate. OCFO, as the business sponsor for its systems, performed the assessment procedures necessary to adequately maintain their systems. However, their applications would be at risk since they rely on the top tier controls of the general support system.

For the systems that are outsourced to DOI-NBC, OCFO does not have processes to monitor the adequacy of security controls maintained by the service provider. In the current year, the agency did not know that the service provider had information security issues until they were provided a reasonable assurance letter, dated October 5, 2005. DOI-NBC reported a serious weakness in complying with OMB Circular A-130, *Management of Federal Information Resources.* DOI-NBC also characterized the weakness as an Improvement Act substantial noncompliance. The risks associated with (1) the lack of timely certification and accreditation, (2) delays in self- assessments, (3) the lack of annual contingency planning, and (4) the outsourced systems' substantial noncompliance, introduce an elevated risk to the NRC's information security system.

NRC management discussed these issues during the annual meeting on management controls as considerations for the agency's Integrity Act reportable items. NRC management is reporting the security risks associated with DOI-NBC as a substantial noncompliance with the Improvement Act.

Recommendations

5. The CFO should coordinate with the Office of Information Services and the Executive Director for Operations to keep abreast of progress in implementing the recommendations made in the OIG's report. This awareness will enable the CFO to better plan his information system security needs.

6. The CFO should establish procedures to monitor and participate in customer advisory work groups on information security issues with the service bureau. At a minimum, the CFO should devise a communication process to stay informed about information security testing and the related results.

Financial Controls Over Disbursements

The OCFO's Division of Financial Services develops and administers policies, standards, and procedures for all financial operation activities of the NRC. The Division is comprised of teams which devote their full attention to time and payroll processing, development of payment policy, processing of obligations, authorizing the payment of nonpayroll transactions, and managing approximately 60 percent of the agency's budget through the Central Allowance Team.

The GAO's *Standards for Internal Control in the Federal Government*, state "Internal control should generally be designed to assure that ongoing monitoring occurs in the normal course of operations. . . . It includes regular management and supervisory activities, comparisons, reconciliations, and other actions people take in performing their duties." The Standards also state, "Management's philosophy and operating style also affect the environment. This fact determines the degree of risk the agency is willing to take and management's philosophy toward performance-based management. Further the attitude and philosophy of management toward information systems, accounting, personnel functions, monitoring, and audits and evaluations can have a profound effect on the internal control."

In the current year, we found that the agency made one improper payment in excess of $1 million. The NRC became aware of the error because the rightful vendor, who was not initially paid, contacted the agency to inquire about the status of its payment. An error of this nature could have been prevented by the controls described below:

- The agency has controls to verify the existence of vendors prior to payment. However, the agency does not have verification controls to review the propriety of edits made to the vendor tables. In the present case, two vendors requested changes to their vendor profiles. Items such as vendor name, address, tax identification number, and bank routing numbers are maintained in the vendor profiles.

- In the prior year, the agency indicated that it was adopting OMB's guidelines for drawing on the Central Contractor Registration (CCR) information to verify electronic funds transfer (EFT) information. The agency's system was planned to verify the propriety of payment information between the NRC's vendor tables and the CCR database prior to payment. We understand that the CCR business rules were implemented relying on internal NRC vendor table maintenance controls, rather than CCR validations. NRC has not implemented the secondary review of vendor table changes.

- The agency does not have controls in place for review and approval of high value payments to non-Federal entities. NRC's high value payments range from amounts in excess of $250 thousand to $300 thousand. Payments in the high value category are not reviewed any differently than payments of nominal value. An independent or secondary validation of these amounts would most likely have detected the payment error.

Additional concerns with this improper payment are (1) the OCFO's lack of disclosure and (2) the lack of understanding of the impact of interest owed on the late invoice to the liabilities included on the Balance Sheet. This condition imposes unnecessary risk in the control environment.

Recommendations

7. The CFO should direct an assessment of financial controls over disbursement activities. At a minimum, the assessment should provide for the development and implementation of second party reviews of the propriety and accuracy of edits to vendor tables.

8. The CFO should periodically assess whether CCR data can be used to provide an electronic validation of EFT information against NRC's payment system prior to certifying the payment.

9. The CFO should establish a secondary review of high value payments. The secondary review should be performed by parties that are not involved directly in payment processing.

10. The CFO should reiterate to all agency managers the importance of having full and open discussion about internal control impacting the reliability and completeness of the agency's financial statements. This discussion could be incorporated during the upcoming implementation of OMB Circular A-123.

Status of Prior Year Comments

In the prior year, we included conditions related to User Organization Compensating Controls and Inadequate Acceptance Testing (included in the Fee Billing System condition) in our report. Corrective actions were implemented during the year to close these two conditions. However, conditions related to the fee billing system and monitoring of accounting for internal use software continued in the current fiscal year.

Report on Compliance with Laws and Regulations

We conducted our audit for the year ended September 30, 2005, in accordance with auditing standards generally accepted in the United States of America, the standards applicable to financial audits contained in *Government Auditing Standards* issued by the Comptroller General of the United States, and OMB Bulletin No. 01-02.

NRC management is responsible for complying with laws and regulations applicable to the agency. As part of obtaining reasonable assurance about whether the agency's financial statements are free of material misstatement, we performed tests of its compliance with certain provisions of applicable regulations, noncompliance with which could have a direct and material effect on the determination of financial statement amounts and certain other laws and regulations specified in OMB Bulletin No. 01-02, including the requirements in the Improvement Act. We limited our tests of compliance to these provisions and we did not test compliance with all laws and regulations applicable to NRC. The results of our tests of compliance disclosed noncompliances with laws and regulations that are required to be reported under *Government Auditing Standards,* OMB Bulletin No. 01-02 or under the Improvement Act.

U.S. Department of Energy Expenses

> NRC's principal statements include reimbursable expenses of DOE National Laboratories. For the years ended September 30, 2005, and 2004, NRC's Statements of Net Cost include approximately $68.7 and $77.2 million, respectively, of reimbursed expenses. Our audits included testing these expenses for compliance with laws and regulations applicable to NRC. The work placed with DOE is under the auspices of a Memorandum of Understanding between NRC and DOE. The examination of DOE National Laboratories for compliance with laws and regulations is DOE's responsibility. This responsibility was further clarified by a memorandum of the GAO's Assistant General Counsel, dated March 6, 1995, where he opined that "...DOE's inability to assure that its contractors' costs [National Laboratories] are legal and proper...does not compel a conclusion that NRC has failed to comply with laws and regulations." DOE also has the cognizant responsibility to assure audit resolution and should provide the results of its audits to NRC.

The objective of our audit of the financial statements was not to provide an opinion on overall compliance with such provisions of laws and regulations and, accordingly, we do not express such an opinion.

Our report contains three noncompliances. There is one noncompliance with OMB Circular A-25, relating to the development of Part 170 hourly rates, which was initially reported in 1998. The other two items are substantial noncompliances with the Improvement Act. The Fee Billing System condition reported in FY 2004 continues to exist. In the current year there is an information system security substantial noncompliance related to the Department of Interior's cross-servicing of the agency's general ledger and payroll service applications. The following discussion addresses the noncompliances.

Part 170 Hourly Rates

As previously reported from FYs 1998 through 2004, the Omnibus Budget Reconciliation Act of 1990 (OBRA-90) requires the NRC to recover approximately 100 percent of its budget authority by assessing fees. (In recent years, the recovery percentage has been reduced by 2 percent each year. During FY 2005, the recovery percentage was 90 percent.) Accordingly, NRC assesses two types of fees to its licensees and applicants. One type, specified in 10 CFR Part 171, consists of annual fees assessed to power reactors, materials and other licensees. The other type, specified in 10 CFR Part 170, and authorized by the Independent Offices Appropriation Act of 1952, is assessed for specific licensing actions, inspections, and other services provided to NRC's licensees and applicants.

Each year, the OCFO computes the hourly rates used to charge for Part 170 services. Consistent with OBRA-90, the rates are based on budgetary data and are used to price individually identifiable Part 170 services. NRC developed the FY 1998 and subsequent years' rates using the budgetary basis without validating the fee amounts to the full cost of providing Part 170 services.

During FYs 2004 and 2005, the agency continued to make progress and is presently refining a strategy to address this noncompliance. At year-end the CFO initiated a dialogue with us on their strategy. Initially the agency was pursuing the development of a validation model to compare budget versus cost-based fee calculations. More recently, the agency has devised a proposed cost-based calculation strategy to develop rates in compliance with OMB Circular A-25, *User Charges*. Once the OCFO completes the implementation of this proposed calculation strategy, we will review the resulting calculations or model and the underlying documented assumptions and data sources used in order to verify the reliability and completeness of the results. The audit assessment will also evaluate the adequacy of fee rule changes, if any. We commend the CFO for their continuing efforts to close this comment.

Recommendation

11.   The CFO should continue to pursue the proposed calculation strategy and develop rates in compliance with OMB Circular A-25. OCFO management should inform the Office of Inspector General of the progress and actions taken to correct this condition.

Fee Billing System

In our *Report on the Effectiveness of Internal Control Over Financial Reporting,* we identified Fee Billing System as both a material weakness and an Improvement Act substantial noncompliance. Refer to that report for a detailed discussion of the condition.

Information System-wide Security Controls

In our *Report on the Effectiveness of Internal Control Over Financial Reporting,* we identified the information system-wide security controls as an Improvement Act substantial noncompliance. Refer to that report for a detailed discussion of the condition.

Status of Prior Year Comments

In the prior year, we included a condition related to Fee Recovery From Licensees in our report. Corrective actions were implemented during the year to close this condition. However, the condition related to Part 170 fees and the Improvement Act substantial noncompliance of the Fee Billing System continued in the current fiscal year.

Internal Control Related to Performance Measures

With respect to internal controls related to performance measures described in Chapter 2 of the *FY 2005 Performance and Accountability Report,* the OIG performed those procedures and will address this issue separately. Our procedures were not designed to provide assurance over reported performance measures and, accordingly, we do not provide an opinion on such information.

Consistency of Other Information

Our audit was conducted for the purpose of forming an opinion on the financial statements of NRC taken as a whole. The required supplementary information referred to in the *Management Discussion and Analysis*, Chapter 1 of this *Performance and Accountability Report*, is not a required part of the financial statements but is supplementary information required by OMB Circular A-136. We have applied certain limited procedures which consisted principally of inquiries of management regarding the methods of measurement and presentation of the supplementary information. However, we did not audit the information and express no opinion on it.

The other accompanying information included in Chapter 2 and the appendices to the *Performance and Accountability Report* are required by OMB Circular A-136 and are presented for purposes of additional analysis and are not a required part of the financial statements. Such information has not been subjected to the auditing procedures applied in the audit of the financial statements and, accordingly, we express no opinion on it.

Our audit was conducted for the purpose of forming an opinion on the financial statements of NRC taken as a whole. The required supplementary information, Schedule of Intragovernmental Assets and Liabilities, and the Schedule of Budgetary Resources, included on pages 150 through 151 of this *Performance and Accountability Report*, is not a required part of the financial statements but is supplementary information required by OMB Circular A-136. This information is also presented for purposes of additional analysis of the financial statements rather than to present the budgetary resources of NRC programs. This information has been subjected to the auditing procedures applied in the audit of the financial statements and, in our opinion, is fairly stated in all material respects in relation to the financial statements taken as a whole.

This report is intended solely for the information and use of NRC management, the Inspector General, OMB, GAO, and the Congress and is not intended to be and should not be used by anyone other than these specified parties.

November 4, 2005

R. Navarro & Associates, Inc.

MANAGEMENT'S RESPONSE TO AUDITORS' REPORTS

UNITED STATES
NUCLEAR REGULATORY COMMISSION
WASHINGTON, D.C. 20555-001

November 9, 2005

MEMORANDUM TO: Stephen D. Dingbaum
Assistant Inspector General for Audits

FROM: Jesse L. Funches
Chief Financial Officer

SUBJECT: AUDIT OF THE FY 2005 FINANCIAL STATEMENTS

I have reviewed the audit report of the agency's FY 2005 financial statements.
Our responses to the recommendations follow.

Recommendation 1

The CFO should direct an assessment of all aspects of the Fee Billing System to ensure
that the remediation plan is updated as necessary and implemented in a timely manner
to enhance the controls over fee billing processes.

Response

Agree. By March 2006, the OCFO will assess the operating aspects of the Fee Billing
System that are essential to the internal control over fee billings, including the processes
related to data obtained from feeder systems, to identify cost-effective controls that will
strengthen the completeness and reliability of the fee billing processes. The remediation
plan will be updated as necessary based on the results of the assessment.

Recommendation 2

The CFO should define, design, and implement compensating controls over the Fee
Billing System, while the system is being considered for redesign.

Response

Agree. By July 2006, the OCFO will use the results of the assessment performed in the response to Recommendation 1, and the experience gained implementing improved controls and quality assurance procedures, to establish additional cost-effective compensating controls in the existing fee billing processes.

Recommendation 3

As the OCFO identifies needed improvement of internal controls outside OCFO's business domain, there should be coordination and collaboration with the Executive Director for Operations as to how, when and to what extent the internal controls should be strengthened in operational program feeder systems, relied upon by OCFO for billing preparation purposes.

Response

Agree.

Recommendation 4

The CFO should continue to promote strengthening internal use software practices, in order to encourage project managers to comply with procedures in effect governing the completeness of new and existing development initiatives.

Response

Agree. The OIG has acknowledged that a significant amount of work has been done by the OCFO over the years to promote and strengthen internal use software practices. However, the OCFO will seek to identify further measures that can be taken and will develop a plan of additional actions by January 2006.

Recommendation 5

The CFO should coordinate with the Office of Information Services and the Executive Director for Operations to keep abreast of progress in implementing the recommendations made in the OIG's report. This awareness will enable the CFO to better plan his information system security needs.

Response

Agree.

Recommendation 6

The CFO should establish procedures to monitor and participate in customer advisory work groups on information security issues with the service bureau. At a minimum, the CFO should devise a communication process to stay informed about information security testing and the related results.

Response

Agree. As a followup to his verbal communications with the service bureau, I will write to the Director and CFO of the Department of Interior National Business Center (DOI-NBC) to document concerns about the security of DOI-NBC's network and mainframe, and to establish an agreement to be informed timely of information security issues and documented completion of corrective actions. The letter will be sent by December 2005.

Recommendation 7

The CFO should direct an assessment of financial controls over disbursement activities. At a minimum, the assessment should provide for the development and implementation of second party reviews of the propriety and accuracy of edits to vendor tables.

Response

Agree. The OCFO will perform an assessment of the financial controls over disbursement activities and make necessary revisions to our existing procedures. This will be completed by June 30, 2006. Additionally, by December 15, 2005, secondary reviews of revisions to the vendor tables in the core financial system will be implemented.

Recommendation 8

The CFO should periodically assess whether, CCR data can be used to provide an electronic validation of EFT information against NRC's payment system prior to certifying the payment.

Response

Agree. The CCR data will be made available in the NRC's core financial system by the end of February 2006, for validation of payments. OCFO will semiannually assess whether CCR data can be used to provide an electronic validation of EFT information for payments. We will perform the next assessment by March 30, 2006.

Recommendation 9

The CFO should establish a secondary review of high value payments. The secondary review should be performed by parties that are not involved directly in payment processing.

Response

Agree. By December 15, 2005, the OCFO will implement secondary reviews of high value payments. These reviews will be performed by staff not involved directly in payment processing.

Recommendation 10

The CFO should reiterate to all agency managers the importance of having full and open discussion about internal control impacting the reliability and completeness of the agency's financial statements. This discussion could be incorporated during the upcoming implementation of OMB Circular A-123.

Response

Agree. I am committed to ensuring that integrity and ethical values are not compromised and will continue to emphasize the importance of having full and open discussion relating to internal controls impacting the reliability and completeness of NRC's financial statements. By December 15, 2005, I will issue a memorandum to all agency managers emphasizing the importance of a full and open discussion relating to internal controls impacting the reliability and completeness of NRC's financial statements.

Recommendation 11

The CFO should continue to pursue the proposed calculation strategy and develop rates in compliance with OMB Circular A-25. OCFO management should inform the Office of Inspector General of the progress and actions taken to correct this condition.

Response

Agree. We will continue with plans to calculate 10 CFR Part 170 hourly rates using actual cost data from the Cost Accounting System. We may use these rates to recover the costs of activities under 10 CFR Part 170 beginning with the FY 2006 fee rule. We will continue to keep the Office of the Inspector General informed of progress and actions.

PRINCIPAL STATEMENTS

BALANCE SHEET
(Dollars in Thousands)

As of September 30,		2005		2004
Assets				
Intragovernmental				
Fund balances with Treasury (Note 2)	$	**220,695**	$	200,277
Accounts receivable (Note 3)		**3,227**		3,357
Other		**1,961**		2,295
Total intragovernmental		**225,883**		205,929
Accounts receivable, net (Note 3)		**60,757**		50,648
Property and equipment, net (Note 4)		**26,983**		26,800
Other		**66**		29
Total Assets	$	**313,689**	$	283,406
Liabilities				
Intragovernmental				
Accounts payable	$	**7,730**	$	8,564
Other (Note 5)		**69,495**		61,568
Total intragovernmental		**77,225**		70,132
Accounts payable		**21,296**		19,367
Federal employees' benefits (Note 6)		**8,417**		8,114
Other liabilities (Note 5)		**49,268**		48,317
Total Liabilities		**156,206**		145,930
Net Position				
Unexpended appropriations		**170,836**		149,901
Cumulative results of operations (Note 8)		**(13,353)**		(12,425)
Total Net Position		**157,483**		137,476
Total Liabilities and Net Position	$	**313,689**	$	283,406

The accompanying notes to the principal statements are an integral part of this statement.

STATEMENT OF NET COST
(Dollars in Thousands)

For the years ended September 30, (Note 9)	2005	2004
Nuclear Reactor Safety		
Gross costs	$ **476,481**	$ 466,440
Less: Earned revenue	**(476,020)**	(478,488)
Total Net Cost of Nuclear Reactor Safety	**461**	(12,048)
Nuclear Materials and Waste Safety		
Gross costs	**206,518**	196,078
Less: Earned revenue	**(73,972)**	(73,679)
Total Net Cost of Nuclear Materials and Waste Safety	**132,546**	122,399
Net Cost of Operations	$ **133,007**	$ 110,351

The accompanying notes to the principal statements are an integral part of this statement.

STATEMENT OF CHANGES IN NET POSITION
(Dollars in Thousands)

For the years ended September 30,	2005		2004	
	Cumulative Results of Operations	Unexpended Appropriations	Cumulative Results of Operations	Unexpended Appropriations
Beginning Balances	$ (12,425)	$ 149,901	$ (8,089)	$ 149,719
Budgetary Financing Sources				
Appropriations received	-	601,245	-	593,000
Appropriations transferred-in/out	-	(463,729)	-	(510,439)
Other adjustments	-	(481)	-	(280)
Appropriations used	116,100	(116,100)	82,099	(82,099)
Non-exchange revenue	7,344	-	725	-
Transfers-in/out without reimbursement	(7,344)	-	(725)	-
Other Financing Sources				
Imputed financing from costs absorbed by others	25,904	-	25,129	-
Other	(9,925)	-	(1,213)	-
Total Financing Sources	132,079	20,935	106,015	182
Net Cost of Operations	(133,007)	-	(110,351)	-
Ending Balances	$ (13,353)	$ 170,836	$ (12,425)	$ 149,901

The accompanying notes to the principal statements are an integral part of this statement.

STATEMENT OF BUDGETARY RESOURCES
(Dollars in Thousands)

For the years ended September 30,	2005	2004
Budgetary Resources		
Budget authority		
Appropriations received	$ 601,245	$ 593,000
Net transfers	68,498	32,905
Unobligated balances		
Beginning of period	36,328	40,572
Spending authority from offsetting collections		
Reimbursements earned	5,836	5,491
Change in unfilled customer orders	431	1,298
Total Spending Authority from Offsetting Collections	6,267	6,789
Recoveries of prior-year obligations	11,019	8,618
Permanently not available	(481)	(280)
Total Budgetary Resources	$ 722,876	$ 681,604
Status of Budgetary Resources		
Obligations incurred (Note 13)		
Direct	$ 659,530	$ 639,322
Reimbursable	6,002	5,953
Unobligated balance		
Apportioned	33,620	35,282
Exempt from apportionment	23,724	1,047
Total Status of Budgetary Resources	$ 722,876	$ 681,604
Relationship of Obligations to Outlays		
Obligated balance, net, beginning of period	$ 157,218	$ 143,934
Obligated balance, net, end of period		
Accounts receivable	(322)	(275)
Unfilled customer orders from Federal sources	(3,885)	(3,882)
Undelivered orders	118,580	117,150
Accounts payable	45,919	44,225
Obligated balance, net, end of period	$ 160,292	$ 157,218
Outlays		
Disbursements	$ 651,389	$ 623,131
Collections	(6,216)	(6,546)
Subtotal	645,173	616,585
Less: Offsetting Receipts	(534,119)	(545,302)
Net Outlays	$ 111,054	$ 71,283

The accompanying notes to the principal statements are an integral part of this statement.

STATEMENT OF FINANCING
(Dollars in Thousands)

For the years ended September 30,	2005	2004
Resources Used to Finance Activities		
Budgetary Resources Obligated		
Obligations incurred (Note 13)	$ 665,532	$ 645,275
Less: Spending authority from offsetting collections and recoveries	(17,286)	(15,407)
Obligations Net of Offsetting Collections and Recoveries	648,246	629,868
Less: Offsetting receipts	(534,119)	(545,302)
Net Obligations	114,127	84,566
Other Resources		
Imputed financing from costs absorbed by others	25,904	25,129
Allocation transfer	2,124	3,206
Other	(9,925)	(1,213)
Net Other Resources Used to Finance Activities	18,103	27,122
Total Resources Used to Finance Activities	132,230	111,688
Resources Used to Finance Items not Part of the Net Cost of Operations		
Change in budgetary resources obligated for goods, services and benefits ordered but not yet provided	(151)	(5,074)
Resources that finance the acquisition of assets	(7,393)	(5,796)
Other	(46)	(217)
Total Resources Used to Finance Items not Part of the Net Cost of Operations	(7,590)	(11,087)
Total Resources Used to Finance the Net Cost of Operations	124,640	100,601
Components of the Net Cost of Operations that will not Require or Generate Resources in the Current Period		
Components Requiring or Generating Resources in the Future Periods		
Increase in annual leave liability	755	2,170
Increase (Decrease) Actuarial Workers' Compensation	303	(959)
Increase (Decrease) in Unfunded Workers' Compensation	228	10
Increase in Unfunded Unemployment	(12)	(5)
Total Components of Net Cost of Operations that will Require or Generate Resources in Future Periods	1,274	1,216
Components not Requiring or Generating Resources:		
Depreciation and amortization	7 093	8,534
Total Components not Requiring or Generating Resources	7,093	8,534
Total Components of Net Cost of Operations that will not Require or Generate Resources in the Current Period	8,367	9,750
Net Costs of Operations	$ 133,007	$ 110,351

The accompanying notes to the principal statements are an integral part of this statement.

NOTES TO PRINCIPAL STATEMENTS

NOTE 1.   SUMMARY OF SIGNIFICANT ACCOUNTING POLICIES

A.   *Reporting Entity*

The NRC is an independent regulatory agency of the Federal Government that was created by the U.S. Congress to regulate the Nation's civilian use of byproduct, source, and special nuclear materials to ensure adequate protection of the public health and safety, to promote the common defense and security, and to protect the environment. Its purposes are defined by the Energy Reorganization Act of 1974, as amended, along with the Atomic Energy Act of 1954, as amended, which provide the foundation for regulating the Nation's civilian use of nuclear materials.

The NRC operates through the execution of its congressionally approved appropriations for salaries and expenses and the Inspector General, including funds derived from the Nuclear Waste Fund. In addition, transfer appropriations are provided by the U.S. Agency for International Development for the development of nuclear safety and regulatory authorities in Russia, Ukraine, Kazakhstan, and Armenia for the independent oversight of nuclear reactors in these countries.

B.   *Basis of Presentation*

These principal statements were prepared to report the financial position and results of operations of the NRC as required by the Chief Financial Officers Act of 1990 and the Government Management Reform Act of 1994. These financial statements were prepared from the books and records of the NRC in conformity with accounting principles generally accepted in the United States of America, the requirements of OMB Circular A-136, *Financial Reporting Requirements*, and NRC accounting policies. These statements are, therefore, different from the financial reports, also prepared by the NRC pursuant to OMB directives, which are used to monitor and control NRC's use of budgetary resources.

NRC has not presented a Statement of Custodial Activity because the amounts involved are immaterial and incidental to its operations and mission.

NRC reclassified the FY 2004 Statement of Net Cost to reflect a change from four strategic arenas to two programs. The two programs are Nuclear Reactor Safety and the Nuclear Materials and Waste Safety. The Nuclear Reactor Safety program is the combination of the FY 2004 Nuclear Reactor Safety and the International Nuclear Safety Support. The Nuclear Materials and Waste Safety program is the combination of FY 2004 Nuclear Materials Safety and Nuclear Waste Safety.

The programs as presented on the Statement of Net Cost are based on the strategic plans and are described as follows:

> **Nuclear Reactor Safety** encompasses all NRC efforts to ensure that civilian nuclear power reactor facilities and research and test reactors are licensed and operated in a manner that adequately protects the public health and safety, and the environment and protects against radiological sabotage and theft or diversion of special nuclear materials. The Nuclear Reactor Safety program contains two activities—Nuclear Reactor Licensing and Nuclear Reactor Inspection.

> **Nuclear Materials and Waste Safety** encompasses all NRC efforts to protect the public health and safety and the environment and ensures the secure use and management of radioactive materials. The Nuclear Materials and Waste Safety program contains five activities—Fuel Facilities Licensing and Inspection, Nuclear Materials Users Licensing and Inspection, High-Level Waste Repository, Decommissioning and Low-Level Waste, and Spent Fuel Storage and Transportation Licensing and Inspection.

C.   *Budgets and Budgetary Accounting*

Budgetary accounting measures appropriation and consumption of budget/spending authority or other budgetary resources and facilitates compliance with legal constraints and controls over the use of Federal funds. Under budgetary reporting principles, budgetary resources are consumed at the time of purchase. Assets and liabilities, which do not consume current budgetary resources, are not reported, and only those liabilities for which valid obligations have been established are considered to consume budgetary resources.

For the past 31 years, Congress has enacted no-year appropriations, which are available for obligation by NRC until expended. The Energy and Water Development Appropriations Act, 2005, requires the NRC to recover approximately 90 percent of its new budget authority of $669.3 million by assessing fees less amounts derived from the Nuclear Waste Fund of $68.5 million. The $669.3 million includes rescissions of $481 thousand to NRC's appropriation from P.L. 108-447 and $552 thousand to the Nuclear Waste Fund appropriation. The $669.3 million does not include any amounts transferred from the U.S. Agency for International Development.

For FY 2004, NRC recovered approximately 92 percent of its new budget authority of $625.6 million less amounts derived from the Nuclear Waste Funds of $32.9 million.

D. Basis of Accounting

Transactions are recorded on an accrual accounting basis. Under the accrual method, revenues are recognized when earned and expenses are recognized when a liability is incurred, without regard to receipt or payment of cash. Interest on borrowings of the U.S. Treasury is not included as a cost to NRC's programs and is not included in the accompanying financial statements.

E. Revenues and Other Financing Sources

The NRC is required to offset its appropriations by the amount of revenues received during the fiscal year by assessing fees. The NRC assesses two types of fees to recover its budget authority: (1) fees assessed under 10 CFR Part 170 for licensing, inspection, and other services under the authority of the Independent Offices Appropriation Act of 1952 to recover the NRC's costs of providing individually identifiable services to specific applicants and licensees; and (2) annual fees assessed for nuclear facilities and materials licensees under 10 CFR Part 171. All fees, with the exception of civil penalties, are exchange revenues in accordance with Statement of Federal Financial Accounting Standards No. 7, Accounting for Revenue and Other Financing Sources and Concepts for Reconciling Budgetary and Financial Accounting.

For accounting purposes, appropriations are recognized as financing sources (appropriations used) at the time expenses are accrued. At the end of the fiscal year, appropriations recognized are reduced by the amount of assessed fees collected during the fiscal year to the extent of new budget authority for the year. Collections which exceed the new budget authority are held to offset subsequent years' appropriations. Appropriations expended for property and equipment are recognized as expenses when the asset is consumed in operations (depreciation and amortization).

F. *Fund Balances with Treasury*

The NRC's cash receipts and disbursements are processed by the U.S. Treasury. The fund balances with the U.S. Treasury are primarily appropriated funds that are available to pay current liabilities and to finance authorized purchase commitments. Funds with Treasury represent NRC's right to draw on the U.S. Treasury for allowable expenditures. All amounts are available to NRC for current use.

G. *Accounts Receivable*

Accounts receivable consist of amounts owed to the NRC by other Federal agencies and the public. Amounts due from the public are presented net of an allowance for uncollectible accounts. The allowance is based on an analysis of the outstanding balances. Receivables from Federal agencies are expected to be collected; therefore, there is no allowance for uncollectible accounts.

H. *Non-Entity Assets*

Accounts receivable include nonentity assets of $5 thousand and $7 thousand at September 30, 2005, and 2004, respectively, and consist of miscellaneous penalties and interest due from the public, which, when collected, must be transferred to the U.S. Treasury.

I. *Property and Equipment*

Property and equipment consist primarily of typical office furnishings, nuclear reactor simulators, and computer hardware and software. The costs of internal use software include the full cost of salaries and benefits from agency personnel involved in software development. The agency has no real property. The land and buildings in which NRC operates are provided by the General Services Administration (GSA), which charges NRC rent that approximates the commercial rental rates for similar properties.

Property with a cost of $50 thousand or more per unit and a useful life of 2 years or more is capitalized at cost and depreciated using the straight-line method over the useful life. Other property items are expensed when purchased. Normal repairs and maintenance are charged to expense as incurred.

J.   *Accounts Payable*

Accounts payable represent vendor invoices for services received by NRC that will be paid at a later date.

K.   *Liabilities Not Covered by Budgetary Resources*

Liabilities represent the amount of monies or other resources that are likely to be paid by NRC as the result of a transaction or event that has already occurred. No liability can be paid by NRC absent an appropriation. Liabilities for which an appropriation has not been enacted and for which there is no certainty that an appropriation will be enacted are classified as Liabilities Not Covered by Budgetary Resources. Also, NRC liabilities arising from sources other than contracts can be abrogated by the Government acting in its sovereign capacity.

Intragovernmental

The U.S. Department of Labor (DOL) paid Federal Employees Compensation Act (FECA) benefits on behalf of NRC which had not been billed or paid by NRC as of September 30, 2005, and 2004, respectively.

Federal Employee Benefits

Federal employee benefits represent the actuarial liability for estimated future FECA disability benefits. The future workers' compensation estimate was generated by DOL from an application of actuarial procedures developed to estimate the liability for FECA, which includes the expected liability for death, disability, medical, and miscellaneous costs for approved compensation cases. The liability was calculated using historical benefit payment patterns related to a specific incurred period to predict the ultimate payments related to that period. These projected annual benefit payments were discounted to present value. The interest rate assumptions utilized for discounting benefits were 4.53 percent for FY 2005 and 4.88 percent for FY 2004.

Other

Accrued annual leave represents the amount of annual leave earned by NRC employees but not yet taken.

L.   *Contingencies*

Contingent liabilities are those where the existence or amount of the liability cannot be determined with certainty pending the outcome of future events. The NRC is a party to various administrative proceedings, legal actions, environmental suits, and claims brought by or against it. The NRC is a party to a case where an adverse outcome is reasonably possible and may exceed $1.3 million. Based on the advice of legal counsel concerning contingencies, it is the opinion of management that the ultimate resolution of these proceedings, actions, suits, and claims will not materially affect the agency's financial statements.

M.   *Annual, Sick, and Other Leave*

Annual leave is accrued as it is earned and the accrual is reduced as leave is taken. Each year, the balance in the accrued annual leave liability account is adjusted to reflect current pay rates. To the extent that current or prior year funding is not available to cover annual leave earned but not taken, funding will be obtained from future financing sources. Sick leave and other types of nonvested leave are expensed as taken.

N.   *Retirement Plans*

NRC employees belong to either the Federal Employees Retirement System (FERS) or the Civil Service Retirement System (CSRS). For FY 2005 and FY 2004, employees belonging to FERS, the NRC withheld 0.8 percent of base pay earnings, in addition to Federal Insurance Contribution Act (FICA) withholdings, and matched the withholdings with an 11.2 percent contribution in FY 2005 and a 10.7 percent contribution in FY 2004. The sum is transferred to the Federal Employees Retirement Fund. For employees covered by CSRS, NRC withholds 7 percent of base pay earnings. The NRC matched this withholding with a 7 percent contribution in FY 2005 and FY 2004.

The Thrift Savings Plan (TSP) is a retirement savings and investment plan for employees belonging to either FERS or CSRS. For employees belonging to FERS, NRC automatically contributes 1 percent of base pay to their account and matches contributions up to an additional 4 percent. The maximum percentage of base pay that an employee participating in FERS may contribute is 15 percent in calendar year 2005, and 14 percent in 2004. Employees belonging to CSRS may contribute up to 10 percent of their salary in calendar year 2005, and 9 percent in 2004, but there is no NRC matching of the contribution. The maximum amount that either FERS or CSRS employees may contribute to the plan is $14 thousand in 2005 and $13 thousand in 2004. The sum of the employees' and NRC's contributions are transferred to the Federal Retirement Thrift Investment Board.

The NRC does not report on its financial statements FERS and CSRS assets, accumulated plan benefits, or unfunded liabilities, if any, applicable to its employees. Reporting such amounts is the responsibility of the U.S. Office of Personnel Management. The portion of the current and estimated future outlays for CSRS not paid by NRC is, in accordance with Statement of Federal Financial Accounting Standards No. 5, Accounting for Liabilities of the Federal Government, included in NRC's financial statements as an imputed financing source.

O.   Leases

The total capital lease liability is funded on an annual basis and included in NRC's annual budget. The NRC's capital leases are for personal property consisting of reproduction equipment which is installed at NRC headquarters. For FY 2005 and FY 2004, there were two capital leases with terms of 5 years each and the interest rate was 4.38 percent for both leases. During FY 2004, a capital lease for a term of 3 years at an interest rate of 6.59 percent was completed. The reproduction equipment is depreciated over 5 years using the straight-line method with no salvage value.

Operating leases consist of real property leases with GSA. The leases are for NRC's headquarters and regional offices. The GSA charges NRC lease rates which approximate commercial rates for comparable space.

P.   U.S. Department of Energy Charges

Financial transactions between the DOE and NRC are fully automated through the U.S. Treasury's Intragovernmental Payment and Collection (IPAC) System. The IPAC System allows DOE to collect amounts due from NRC directly from NRC's account at the U.S. Treasury for goods and/or services rendered. Project manager verification of goods and/or services received is subsequently accomplished through a system-generated voucher approval process. The vouchers are returned to the Office of the Chief Financial Officer documenting that the charges have been accepted.

Q.   *Pricing Policy*

The NRC provides goods and services to the public and other Government entities. In accordance with OMB Circular No. A-25, User Charges, and the Independent Offices Appropriation Act of 1952, NRC assesses fees under 10 CFR Part 170 for licensing and inspection activities to recover the full cost of providing individually identifiable services.

The NRC's policy is to recover the full cost of goods and services provided to other Government entities where (1) the services performed are not part of its statutory mission and (2) NRC has not received appropriations for those services. Fees for reimbursable work are assessed at the 10 CFR Part 170 rate with minor exceptions for programs that are nominal activities of the NRC.

R.   *Net Position*

The NRC's net position consists of unexpended appropriations and cumulative results of operations. Unexpended appropriations represent appropriated spending authority that is unobligated and has not been withdrawn by the U.S. Treasury, and obligations that have not been paid. Cumulative results of operations represent the excess of financing sources over expenses since inception.

S.   *Use of Management Estimates*

The preparation of the accompanying financial statements in accordance with generally accepted accounting principles requires management to make certain estimates and assumptions that directly affect the results of reported assets, liabilities, revenues, and expenses. Actual results could differ from these estimates.

NOTE 2.   FUND BALANCES WITH TREASURY

(In thousands)	2005	2004
Fund Balances		
Appropriated funds	$ **217,637**	$ 193,547
Allocation transfers	**3,047**	3,839
Other fund types	**11**	2,891
Total	$ **220,695**	$ 200,277
Status of Fund Balance with Treasury		
Unobligated Balance		
Available		
Appropriated funds	$ **57,344**	$ 36,329
Allocation transfers	**1,638**	1,857
Unavailable	**155**	3,046
Obligated balance not yet disbursed	**161,558**	159,045
Total	$ **220,695**	$ 200,277

The Fund Balance with Treasury consists of Unobligated and Obligated Balances budgetary accounts. It includes Nuclear Waste Fund activity. The Nuclear Waste Fund Unobligated balance is $23.7 million as of September 30, 2005, and $1.0 million as of September 30, 2004.

NOTE 3.   ACCOUNTS RECEIVABLE

(In thousands)	2005	2004
Intragovernmental		
Receivables and reimbursements	$ 3,227	$ 3,357
Receivables with the Public		
Materials and facilities fees - billed	$ 2,124	$ 3,060
Materials and facilities fees - unbilled	61,482	49,684
Other	38	37
Total Accounts Receivable	63,644	52,781
Less: Allowance for uncollectible accounts	(2,887)	(2,133)
Accounts Receivable, Net	$ 60,757	$ 50,648

NOTE 4.   PROPERTY AND EQUIPMENT, NET

(In thousands)

Fixed Assets Class	Service Years	Acquisition Value	Accumulated Depreciation and Amortization	2005 Net Book Value	2004 Net Book Value
Equipment	5-8	$ 15,297	$ (13,931)	$ 1,366	$ 1,824
IT software	5	41,238	(35,420)	5,818	10,288
IT software under development	-	8,303	-	8,303	4,014
Leasehold improvements	20	21,857	(12,439)	9,418	10,617
Leasehold improvements in progress	-	2,078	-	2,078	57
		$ 88,773	$ (61,790)	$ 26,983	$ 26,800

NOTE 5.   OTHER LIABILITIES

(In thousands)	2005	2004
Intragovernmental		
Liability to offset net accounts receivable for fees assessed	$ 63,627	$ 53,704
Liability from fees collected which will offset current year's appropriations	6	2,857
Liability to offset miscellaneous accounts receivable	3	7
Liability for advances from other agencies	2,090	1,778
Accrued workers' compensation	1,883	1,655
Accrued unemployment compensation	12	24
Employee benefit contributions	1,874	1,543
Total Intragovernmental Other Liabilities	$ 69,495	$ 61,568

The liability to offset the net accounts receivable for fees assessed represents amounts which, when collected, will be transferred to the U.S. Treasury to offset NRC's appropriations in the year collected.

(In thousands)	2005	2004
Accrued annual leave	$ 32,960	$ 32,205
Accrued salaries	12,986	13,001
Contract holdbacks, advances, and other	3,322	3,111
Total Other Liabilities	$ 49,268	$ 48,317

Other liabilities, except accrued annual leave, contract holdbacks, and advances from others, are current.

NOTE 6.   LIABILITIES NOT COVERED BY BUDGETARY RESOURCES

(In thousands)	2005	2004
Intragovernmental		
FECA paid by DOL	$ 1,883	$ 1,655
Accrued unemployment compensation	12	24
Federal Employee Benefits		
Future FECA	8,417	8,114
Other		
Accrued annual leave	32,960	32,205
Total Liabilities not Covered by Budgetary Resources	$ 43,272	$ 41,998

Balance Sheet amounts represent ending balances of worker's compensation and annual leave. The Statement of Financing amount represents the change in activity in worker's compensation and annual leave balance.

NOTE 7.   LEASES

(In thousands)

Future Lease Payments Due:

Fiscal Year	Capital	Operating	2005	2004
2005	$ -	$ -	$ -	$ 22,980
2006	164	22,538	**22,702**	22,218
2007	128	22,627	**22,755**	22,271
2008	-	21,496	**21,496**	21,013
2009	-	19,347	**19,347**	18,864
2010 and thereafter	-	89,288	**89,288**	88,167
Total	292	175,296	**175,588**	195,513
Add: imputed interest	12	-	**12**	28
Total Future Lease Payments	$ 304	$ 175,296	**$ 175,600**	$ 195,541

NOTE 8.   CUMULATIVE RESULTS OF OPERATIONS

(In thousands)	2005		2004	
Future funding requirements	$	(43,272)	$	(41,998)
Investment in property and equipment, net		26,983		26,800
Contributions from foreign cooperative research agreements		2,867		2,739
Other		69		34
Cumulative Results of Operations	$	(13,353)	$	(12,425)

Future funding requirements represent the amount of future funding needed to pay the accrued unfunded expenses as of September 30, 2005, and 2004. These accruals are not funded from current or prior-year appropriations and assessments, but rather should be funded from future appropriations and assessments. Accordingly, future funding requirements have been recognized for the expenses that will be paid from future appropriations.

NOTE 9.   STATEMENT OF NET COST

(In thousands)

For the years ended September 30,	2005		2004	
Nuclear Reactor Safety				
Intragovernmental gross costs	$	143,035	$	145,494
Less: Intragovernmental earned revenue		(29,299)		(28,829)
Intragovernmental net costs		113,736		116,665
Gross costs with the public		333,446		320,946
Less: Earned revenue from the public		(446,721)		(449,659)
Net costs with the public		(113,275)		(128,713)
Total Net Cost of Nuclear Reactor Safety	$	461	$	(12,048)
Nuclear Materials and Waste Safety				
Intragovernmental gross costs	$	48,551	$	51,832
Less: Intragovernmental earned revenue		(5,113)		(5,085)
Intragovernmental net costs		43,438		46,747
Gross costs with the public		157,967		144,246
Less: Earned revenue from the public		(68,859)		(68,594)
Net costs with the public		89,108		75,652
Total Net Cost of Nuclear Materials and Waste Safety	$	132,546	$	122,399

For "Intragovernmental gross costs," the buyers and sellers are both Federal entities. For "Earned revenues from the public," the buyers of the goods or services are non-Federal entities.

NOTE 10.   EXCHANGE REVENUES

(In thousands)	2005	2004
Fees for licensing, inspection, and other services	$ **544,044**	$ 546,515
Revenue from reimbursable work	**5,948**	5,652
Total Exchange Revenues	$ **549,992**	$ 552,167

NOTE 11.   BUDGET FUNCTIONAL CLASSIFICATION

(In thousands)			2005	2004
		Earned		
Functional Classification	*Gross Cost*	*Revenue*	*Net Cost*	*Net Cost*
276 - Energy Information, Policy, & Regulation	$ 680,350	$ 549,992	$ **130,358**	$ 107,105
150 - AID International Affairs	2,649	-	**2,649**	3,246
Total	$ 682,999	$ 549,992	$ **133,007**	$ 110,351

Intragovernmental

			2005	2004
		Earned		
Functional Classification	*Gross Cost*	*Revenue*	*Net Cost*	*Net Cost*
276 - Energy Information, Policy, & Regulation	$ 188,937	$ 34,412	$ **154,525**	$ 160,165
150 - AID International Affairs	2,649	-	**2,649**	3,246
Total	$ 191,586	$ 34,412	$ **157,174**	$ 163,411

NOTE 12.   FINANCING SOURCES OTHER THAN EXCHANGE REVENUE

(In thousands)

Appropriated Funds Used

Collections were used to reduce the fiscal year's appropriations recognized:

	2005	2004
Funds consumed	$ 650,219	$ 627,401
Less: collection from fees assessed	(534,119)	(545,302)
Appropriated funds used	$ 116,100	$ 82,099

Funds consumed include $38.3 million and $43.8 million through September 30, 2005, and 2004, respectively, of available funds from prior years.

Non-Exchange Revenue

	2005	2004
Civil penalties	$ 5,807	$ 622
Miscellaneous receipts	1,537	103
Total Non-Exchange Revenue	$ 7,344	$ 725

Imputed Financing

	2005	2004
Civil Service Retirement System	$ 11,993	$ 13,073
Federal Employee Health Benefit	13,735	11,924
Federal Employee Group Life Insurance	62	57
Judgements Awards	114	75
Total Imputed Financing	$ 25,904	$ 25,129

Transfers In/Out

	2005	2004
Transfers out to Treasury		
License fees	$ 534,119	$ 545,302
Non-exchange revenue	7,344	725
Total Transfers-Out to Treasury	$ 541,463	$ 546,027

NOTE 13.   TOTAL OBLIGATIONS INCURRED

(In thousands)

	2005	2004
Direct Obligations		
Category A	$ 613,502	$ 606,764
Exempt from Apportionment	46,028	32,558
Total Direct Obligations	659,530	639,322
Reimbursable Obligations	6,002	5,953
Total Obligations Incurred	$ 665,532	$ 645,275

Obligations exempt from apportionment are the result of funds derived from the Nuclear Waste Fund. Category A Obligations consist of NRC appropriations only.

NOTE 14.   NUCLEAR WASTE FUND

Included in NRC's budget for FY 2005 and 2004 are $68.5 million and $32.9 million, respectively, provided from the Nuclear Waste Fund. In accordance with Statement of Federal Financial Accounting Standards No. 27, Identifying and Reporting Earmarked Funds, NRC has determined that funding from the Nuclear Waste Fund does not fully meet the definition as an earmarked fund. However, in order to provide additional information to the users of these financial statements, enhanced disclosure of the fund is presented below.

The Nuclear Waste Fund was authorized by the Nuclear Waste Policy Act of 1982 (PL 97-425). The funding provided to NRC in FY 2005 and 2004 for the purpose of performing activities associated with the DOE's application for a high-level waste repository at Yucca Mountain, Nevada. These activities included assistance to DOE with the application, review of the application, the conduct of thorough safety and security evaluations, preparation of the safety evaluation report, initiation of the inspection program, ensuring that the regulation process was made available to stakeholders and the general public, and to provide legal advice and representation for staff reviews and Commission actions.

The Nuclear Waste Fund amounts received, expended, obligated, and unobligated balances as of September 30, 2005, and 2004, are shown in the following:

(In thousands)	2005	2004
Appropriations Received	$ 68,498	$ 32,905
Expended Appropriations	41,976	29,985
Obligations Incurred	46,028	32,558
Unobligated Balances	23,724	1,047

NOTE 15.   RECLASSIFICATIONS

Costs incurred in FY 2004 for an IT software project under development have been reclassified in the FY 2004 amounts. Costs of $117 thousand have increased the Property and Equipment, Net, in the Balance Sheet and reduced the gross costs in the Nuclear Reactor Safety program in the Statement of Net Cost. Thus FY 2005 Beginning Cumulative Results of Operations has been adjusted.

The four strategic arenas for the year ended September 30, 2004, were reclassified into the two programs used for the year ended September 30, 2005, presentation. The Net Cost of Operations below shows the reclassification of the 2004 strategic arenas into programs:

		Program Reclassification	
Strategic Arenas	2004	Nuclear Reactor Safety	Nuclear Materials and Waste Safety
Nuclear Reactor Safety	$ (26,457)	$ (26,457)	$ -
Nuclear Materials Safety	40,215	-	40,215
Nuclear Waste Safety	82,184	-	82,184
International	14,409	14,409	-
Total Cost of Operations	$ 110,351	$ (12,048)	$ 122,399

REQUIRED SUPPLEMENTARY INFORMATION

Schedule of Intragovernmental Assets and Liabilities

(Dollars in thousands)

As of September 30,	2005	2004
Intragovernmental Assets		
Fund Balance with Treasury		
Department of the Treasury	$ 220,695	$ 200,277
Accounts Receivable		
Tennessee Valley Authority	2,778	2,952
Department of Energy	271	171
Other Agencies	178	234
Total Accounts Receivable	$ 3,227	$ 3,357
Other Assets		
General Services Administration	254	-
Department of Commerce	472	778
Department of Interior	509	588
Department of the Navy	606	634
Department of Labor	42	42
Department of Energy	20	-
Office of Personnel Management	12	-
Other Agencies	46	253
Total Other Assets	1,961	2,295
Total Intragovernmental Assets	$ 225,883	$ 205,929

As of September 30,	2005	2004
Intragovernmental Liabilities		
Accounts Payable		
General Services Administration	$ 950	$ 1,223
Department of Energy	5,667	6,330
Office of Personnel Management	180	-
Other Agencies	933	1,011
Total Accounts Payable	7,730	8,564
Other Liabilities		
Department of the Treasury - General Fund	63,633	56,561
Department of Labor	1,895	1,679
Office of Personnel Management	1,874	1,543
Other Agencies	2,093	1,785
Total Other Liabilities	69,495	61,568
Total Intragovernmental Liabilities	$ 77,225	$ 70,132

Schedule of Budgetary Resources

(Dollars in thousands)
For the years ended September 30, 2005

	X0200	X0300	Total
Budgetary Resources			
Budget authority			
Appropriations received	$ 593,727	$ 7,518	$ 601,245
Net transfers	68,498	-	68,498
Unobligated balances			
Beginning of period	35,032	1,296	36,328
Spending authority from offsetting collections			
Reimbursements earned	5,836	-	5,836
Change in unfilled customer orders	431	-	431
Total Spending Authority from Offsetting Collections	6,267	-	6,267
Recoveries of prior year obligations	10,799	220	11,019
Permanently not available	(475)	(6)	(481)
Total Budgetary Resources	$ 713,848	$ 9,028	$ 722,876
Status of Budgetary Resources:			
Obligations incurred			
Direct	$ 651,583	$ 7,947	$ 659,530
Reimbursable	6,002	-	6,002
Unobligated balance			
Apportioned	32,539	1,081	33,620
Exempt from apportionment	23,724	-	23,724
Total Status of Budgetary Resources	$ 713,848	$ 9,028	$ 722,876
Relationship of Obligations to Outlays:			
Obligated balance, net, beginning of period	$ 155,918	$ 1,300	$ 157,218
Obligated balance, net, end of period:			
Accounts receivable	(322)	-	(322)
Unfilled customer orders from Federal sources	(3,885)	-	(3,885)
Undelivered orders	117,443	1,137	118,580
Accounts payable	45,580	339	45,919
Obligated balance, net, end of period	$ 158,816	$ 1,476	$ 160,292
Outlays:			
Disbursements	$ 643,837	$ 7,552	$ 651,389
Collections	(6,216)	-	(6,216)
Subtotal	637,621	7,552	645,173
Less: Offsetting receipts	(527,358)	(6,761)	(534,119)
Net Outlays	$ 110,263	$ 791	$ 111,054

Promoting the
security of
our Nation . . .

INSPECTOR GENERAL'S ASSESSMENT OF THE MOST SERIOUS MANAGEMENT CHALLENGES FACING NRC

September 30, 2005

MEMORANDUM TO: Chairman Diaz

FROM: Hubert T. Bell
        Inspector General

SUBJECT: INSPECTOR GENERAL'S ASSESSMENT OF
        THE MOST SERIOUS MANAGEMENT CHALLENGES
        FACING NRC (OIG-05-A-23)

SUMMARY

On January 24, 2000, Congress enacted the *Reports Consolidation Act of 2000* (the Act), which requires Federal agencies to provide financial and performance management information in a more meaningful and useful format for the Congress, the President, and the public. Included in the Act is the requirement that, on an annual basis, the Inspector General of each Federal agency summarize what he or she considers to be the most serious management and performance challenges facing the agency and assess the agency's progress in addressing those challenges. In compliance with the Act, I am submitting my annual assessment of the most serious management challenges confronting the United States Nuclear Regulatory Commission (NRC). Also, included in this submission is a listing of Office of the Inspector General (OIG) audit and investigative reports issued during FY 2005. These reports address the challenges identified.

Congress left the determination and threshold of what constitutes a most serious management challenge to the discretion of the Inspectors General. Therefore, I applied the following definition in preparing my statement:

> Serious management challenges are mission critical areas or programs that have the *potential* for a perennial weakness or vulnerability that, without substantial management attention, would seriously impact agency operations or strategic goals.

The most serious management challenges facing NRC may be, but are not necessarily, areas that are problematic for the agency. The challenges identified represent critical areas or difficult tasks that warrant high-level management attention. This year, I identified nine management challenges that I consider to be the most serious. These challenges are essentially the same ones identified last year, with minor title changes for challenges 3 and 4.

DISCUSSION

The most serious management challenges that follow are not ranked in any order of importance.

CHALLENGE 1

Protection of nuclear material used for civilian purposes.

NRC's Strategic Plan provides for "Excellence in regulating the safe and secure use and management of radioactive materials for the public good." NRC is authorized to grant licenses for the possession and use of radioactive materials (e.g., byproduct material,[a] source material,[b] and special nuclear material[c]) and establish regulations to govern the possession and use of those materials. NRC's regulations require that certain materials licensees have extensive material control and accounting programs as a condition of their license. All other license applicants (including those requesting authorization to possess small quantities of special nuclear materials) must develop and implement plans that demonstrate a commitment to accurately control and account for radioactive materials.

One of NRC's and the nuclear industry's highest priorities must be ensuring adequate protection of public health and safety. Heightened sensitivity to the control of special nuclear materials warrants NRC's serious attention to its licensees' material control and accounting activities. The challenges currently facing NRC will be to (1) ensure that there are adequate inspections to verify licensees' commitments to their material control and accounting programs, and a reliable special nuclear materials system; and (2) establish a means to ensure the accurate accounting for radioactive materials, especially those with the greatest potential to impact public health and safety.

a Byproduct material – (1) Any radioactive material (except special nuclear material) yielded in or made radioactive by exposure to the radiation incident to the process of producing or utilizing special nuclear material and (2) the tailings or wastes produced by the extraction or concentration of uranium or thorium from any ore processed primarily for its source material content. [Source: Atomic Energy Act of 1954, Section 11 (e)]

b Source material – Uranium or thorium or any combination thereof, in any physical or chemical form; or ores that contain by weight 0.05 percent or more of (1) uranium, (2) thorium, or (3) any combination thereof. Source material includes depleted uranium and natural uranium, but not "special nuclear material." [Source: Title 10 *Code of Federal Regulations* (CFR) Part 40.4]

c Special nuclear material – Plutonium, uranium-233, uranium enriched in the isotopes uranium-233 or uranium-235, and any other material which the Commission, pursuant to the provisions of Section 51 of the Atomic Energy Act of 1954, as amended, determines to be special nuclear material, but does not include source material; or any material artificially enriched by any of the foregoing, but does not include source material. [Source: Title 10 CFR Part 74.4]

The NRC has proposed rulemaking for a National Source Tracking System effort to improve accountability and tracking of radioactive sources. The system is proposed as a cradle to grave tracking system of high risk sealed sources. The NRC has also worked with the Department of Energy (DOE) to facilitate recovery of selected orphaned sources. Further, NRC also issued security orders to selected material licensees requiring upgraded physical security.

NRC regulations require stringent design, testing, and monitoring in the handling and storage of spent nuclear fuel. In July 2005, NRC began site-specific spent fuel pool assessments to identify additional enhancements. Nine plant assessments were completed in July and an additional 57 assessments are scheduled for completion by the end of the calendar year.

Related Office of the Inspector General Work

Audits

- Audit of NRC's Baseline Inspection Program
- Audit of NRC's High-Level Waste Program
- Audit of NRC's Generic Communications Program
- Audit of the Decommissioning Program

Investigations

- Review of NRC Oversight of Licensee Fitness for Duty Program
- Inadequate Handling of Inspection Finding
- Adequacy of NRC Oversight of Lost Nuclear Material
- NRC's Oversight of the Hope Creek Nuclear Power Plant

CHALLENGE 2
Protection of information.

Information is an asset and must be protected. Information needing protection includes sensitive unclassified and classified information as well as computer security information.

Sensitive Unclassified and Classified Information

As a result of increased terrorist activity worldwide, NRC continues to reexamine its practice of releasing most documents to the public. NRC employees create and work on significant amounts of information that is sensitive and needs to be protected. Such information can be sensitive unclassified information or classified national security information contained in written documents and various electronic databases.

The agency has made strides in evaluating information that should be withheld from public release. There have been some instances, however, where stakeholders discovered certain documents that should have been publicly available, but instead were designated as non-publicly available. In other instances, sensitive unclassified documents including Safeguards Information (SGI) that should have been withheld were inadvertently released. In light of these occurrences, the agency needs to be particularly vigilant in determining which documents and information should be released.

NRC reviewed other areas to ensure the appropriate release of documents to the public. For example, the Sensitive Information Screening Project (SISP) has provided guidance on the handling of information that may be of benefit to terrorists and the treatment of such information. The agency also developed new guidance for handling sensitive unclassified information in the Agencywide Documents Access and Management System (ADAMS) that could potentially be useful to a terrorist. The Commission approved criteria for staff to use in performing reviews of documents to ensure that only appropriate information is publicly released.

The Executive Director for Operations (EDO) has emphasized the importance of being vigilant about documents sent via e-mail, because the implications of an inadvertent e-mail transmittal of sensitive unclassified or classified information can be enormous. The EDO reminded NRC employees that they should not send sensitive or classified documents through the e-mail system. In addition, on July 29, 2005, NRC issued a policy reminder that camera-equipped cell phones pose a security as well as a privacy concern because they enable people to covertly photograph images or scenes and transmit them instantly to the Internet.

Computer Security

Computer security is the protection afforded to an information system in order to attain the objective of preserving the integrity, availability, and confidentiality of the information system resources (including hardware, software, firmware, information/ data, and telecommunications).

The Federal Information Security Management Act (FISMA) was enacted on December 17, 2002. FISMA outlines the information security management requirements which all agencies must implement and report annually to the Office of Management and Budget (OMB) and Congress on the effectiveness of their security programs. This evaluation must include testing of the effectiveness of information security policies, procedures, and practices of a representative subset of the agency's information systems. In addition to the agency review, FISMA requires an annual evaluation to be performed by the Office of the Inspector General. This year's OIG evaluation discovered weaknesses in the following security controls: certification and accreditation process, automated information system inventory process, security controls for standalone personal computers and laptops, listed systems that process safeguards and/or classified information, and weaknesses in complying with many of the OMB requirements for FISMA implementation.

Related Office of the Inspector General Work

Audits

- Audit of NRC's Drug Testing Program

- Independent Evaluation of NRC's Implementation of the Federal Information Security Management Act (FISMA) for Fiscal Year 2005

- Audit of the Reactor Program System

- Audit of NRC's Telecommunications Program

- Audit of NRC's Policy and Practices Concerning Camera Cell Phones

Investigations

- Failure to Docket Licensee Information

- Follow-up to Inadvertent Releases of SGI to Unauthorized Individuals

- Mishandling of Allegation Regarding Design Basis Information

CHALLENGE 3
Development and implementation of a risk-informed and performance-based regulatory approach.

The Chairman has stated that NRC has increased its safety focus on licensing and oversight activities through application of a balanced combination of experience, deterministic models, and probabilistic analysis. This approach is known as risk-informed and performance-based regulation. However, NRC continues to face different challenges in making its regulatory framework more risk-informed for nuclear power plants and nuclear material licensees. Incorporating risk analysis into regulatory decisions improves the regulatory process by focusing both NRC and licensee attention and activities on the areas of highest risk. The result may be reducing unnecessary burden on licensees and increasing the efficiency and effectiveness of the agency's resources.

NRC conducts inspections at the Nation's 104 nuclear power reactors licensed to operate. The NRC Reactor Inspection Program and Reactor Performance Assessment Program are combined into a single program. This combined program implements the revised reactor oversight process (ROP). An integral part of the ROP is the baseline inspection program that was developed using a risk-informed approach to determine a comprehensive list of areas to inspect within seven established cornerstones of safety. While the baseline inspection program framework is generally sound, OIG identified opportunities for improvement. The baseline inspection program is the minimum inspection oversight that should be conducted at each nuclear power plant, but the agency lacks a mechanism to assess the overall effectiveness and quality of this critical program.

Another challenge is NRC's use of Probabilistic Risk Assessment (PRA). This challenge reflects NRC's commitment to increase the use of PRA technology in all regulatory matters to the extent supported by the state-of-the-art in PRA methods and data and in a manner that complements the agency's approach and philosophy. Implementation of this policy is expected to improve NRC's regulation of licensees.

Related Office of the Inspector General Work

Audits

- Audit of NRC's Baseline Inspection Program
- Audit of NRC's Generic Communications Program

CHALLENGE 4
Ability to modify regulatory processes to meet a changing environment.

As a result of the changing regulatory and business environment, areas of increased emphasis exist. These areas are detailed in the NRC Strategic Plan. External as well as internal demands drive the NRC towards ensuring that it is more open in its regulatory processes. This openness results in a constant balancing of long-term improvement efforts and short-term emergent issues. NRC continues to face challenges related to its ability to address increased workloads associated with reactor license renewals, new plant licensing, licensee requests to increase power levels, and high-level waste disposal.

Reactor License Renewals

NRC's license renewal program is one of the major elements of its regulatory work. In accordance with the Atomic Energy Act, NRC approves and issues licenses for commercial power reactors to operate for up to 40 years. The Act allows the NRC to approve these licenses to be renewed for an additional 20 years. Among the nuclear power plants that have not yet had their license renewed, the first of these 40-year operating licenses will expire in 2009. Approximately 25 percent of the remaining licenses will expire by 2015. The decision whether to seek a renewal is the responsibility of the nuclear power plant owner(s). There continues to be a sustained strong interest in license renewal from utilities. To regulate this activity, NRC reviews the applicants' technical submittals and environmental application materials to verify information submitted in the renewal applications. An application for license renewal addresses technical and environmental issues.

New Plant Licensing

There is a growing list of United States utilities (licensees) that are publicly considering new plant construction in the U.S. NRC's licensing process outlined in 10 CFR Part 52 involves a review of Early Site Permits (ESP), Standard Design Certifications, and/or Combined Operating Licenses (COLs) for nuclear facilities. The COLs for nuclear power facilities involves the issuance of a combined construction permit and a conditional operating license for a nuclear power facility. NRC is involved in several significant activities to ensure that it is prepared to review the first of these COL applications which is expected in 2007-2008. Some of these activities include:

- Reviewing industry's guidelines for a COL application,

- Determining what actions are necessary to prepare for receipt of a COL application,

- Assessing rulemaking activities for the licensing process, and

- Reviewing ESP applications.

Although the Part 52 application process has advantages for both NRC and the nuclear industry, it nevertheless, represents a significant challenge through the increased workload and pressure on the agency to create the infrastructure necessary to support review of new plant licensing applications.

Also, NRC has certified reactor designs, which the agency reviews and approves for general use. Licensees' use of a pre-approved design streamlines and shortens the NRC review process, ultimately paving the way for new reactors to be built and licensed.

Licensee Requests to Increase Power Levels

Many licensees have sought NRC approval to operate their plants at a higher power level than the level authorized in the original license by submitting a request to increase reactor power output. As of April 2005, the NRC approved over 105 power uprate increases. Over the next five years, licensees anticipate requesting additional power uprates, which will affect the ability of NRC staff to maintain established review schedules.

High-Level Waste Disposal

According to the Nuclear Waste Policy Act, the DOE has the responsibility to locate, design, build, and operate a repository for high-level nuclear waste. NRC has the responsibility to license and regulate this facility. Over the past several years, NRC has been preparing its license application review plan. DOE's plans to tender a license application to NRC for the construction of a permanent repository for high-level nuclear waste at Yucca Mountain in Nevada were delayed by a court ruling in FY 2004. The court ruling vacated the Environmental Protection Agency's (EPA) 10,000-year compliance period standard because it was not consistent with the recommendation of the National Academy of Sciences, as mandated by Congress. As a result, EPA needed to develop public health standards for the planned high-level radioactive waste disposal facility at Yucca Mountain that will protect the public health for 1 million years. EPA posted a notice of the proposed standards in the *Federal Register*, established a period for public comment, and will hold public hearings on the proposed rule during the comment period. NRC is revising its regulations in this area, as Congress also mandated that NRC incorporate the EPA standards in its regulations. On September 8, 2005, NRC published a *Federal Register* notice proposing to amend its regulations to implement EPA's proposed standards.

The revised schedule for DOE to tender a license application has not yet been established. Because a multitude of issues will need review in a congressionally mandated 3- to 4-year time frame, NRC anticipates that the administrative proceeding to assess the repository will be an enormous undertaking. One significant challenge for NRC is ensuring that all parties to the licensing process and key decision makers have timely access to filings and exhibits involved with the licensing process.

An additional delay resulted from a ruling by an NRC Atomic Safety Licensing Board Panel that DOE improperly certified that it had met its regulatory obligation to make all of its documentary material related to Yucca Mountain electronically available via the NRC's Licensing Support Network. This ruling was a significant determination, as it is DOE's certification that starts a six-month clock for the earliest that NRC can docket DOE's Yucca Mountain license application. DOE has been working towards re-certifying that all of its documentary material related to Yucca Mountain is electronically available. As of July 2005, over 3 million documents had been loaded into the licensing support network. Recertification has not yet been rescheduled.

Given the events of FYs 2004 and 2005, the ability to modify regulatory processes to meet a changing environment will continue to be a prominent challenge for NRC in FY 2006, as it relates to NRC's high-level waste program.

Related Office of the Inspector General Work

Audits

- Audit of NRC's High-Level Waste Program
- Audit of the Reactor Program System
- Audit of NRC's Generic Communications Program

Investigation

- Adequacy of NRC's Oversight of Vermont Yankee Power Uprate

CHALLENGE 5
Implementation of information resources.

Federal agencies' acquisition and implementation of information resources are crucial to (1) support critical mission-related operations and (2) provide more effective and cost-efficient Government services to the public. The necessary link of information technology (IT) to NRC's mission performance makes it important to have decision-making processes which ensure that funds are invested and managed to achieve high value outcomes at acceptable costs. NRC relies on a wide variety of information systems to help it fulfill its responsibilities and support its business flow. NRC, like other Federal agencies, continues to work towards obtaining a good return on these investments. In recent years, NRC has created large databases of publicly available information, including the High-Level Waste Meta System, the licensing support network (LSN), the NRC Web site, and the ADAMS public reading room.

The following sections highlight NRC's efforts to strengthen and support the agency's business needs using information technology strategies.

Project Management Methodology (PMM)

NRC is subject to several legislative mandates regarding its management of IT investments. The OMB has issued circulars that describe policies to be implemented at each agency. These policies are summarized and referenced in OMB Circular A-130, "Management of Federal Information Resources." In addition, the Clinger-Cohen Act of 1996 requires each Agency head to design and implement a Capital Planning and Investment Control process. The NRC developed PMM to address these requirements.

Homeland Security Presidential Directive 12 (HSPD-12)

On August 27, 2004, the President signed HSPD-12 requiring the development and agency implementation of a mandatory Governmentwide standard for secure and reliable forms of identification for Federal employees and contractors. According to HSPD-12, secure and reliable forms of identification means identification that:

- Is based on sound criteria for verifying an individual employee's identity;

- Is strongly resistant to identity fraud, tampering, counterfeiting, and terrorist exploitation;

- Uses electronic methods of rapid authentication; and

- Is issued only by providers whose reliability has been established by an official accreditation process.

The agency formed an HSPD-12 working group that includes representatives from the Office of Administration's Division of Facilities and Security/Security Branch; Planning, Budgeting, and Performance Management Team; and the Office of Information Services. This working group has been meeting on a bi-weekly basis to discuss the requirements and impact of HSPD-12 and the associated guidance publications.

NRC will have a challenge meeting the completion dates and having the resources to conduct the activities.

High-Level Waste (HLW) Meta System

NRC is developing the High-Level Waste Meta System (a system of systems) to support the agency's review and hearings pertaining to the DOE's anticipated application to build a high-level waste repository at Yucca Mountain. The HLW Meta System is the collection of interdependent software applications, procedures, and information technology needed to support NRC's business activities associated with the licensing process. For example, the system will interface with ADAMS and the licensing support network and will include an Electronic Information Exchange component to allow parties to submit, service, and access documents. It will also include the Electronic Hearing Docket, which will serve as the agency's official docket; the Digital Data Management System, which will submit exhibits and hearing transcripts to support hearing functions; and NRC's High-Level Waste Collection of records relevant to discovery. System development is expected to cost approximately $9.5 million and NRC staff anticipates that much of the system will be certified and accredited by April 2006. The challenge for NRC will be to ensure that this important project stays on track in order to effectively support the upcoming license application review process.

Related Office of the Inspector General Work

Audits

- Independent Evaluation of NRC's Implementation of the Federal

- Information Security Management Act (FISMA) for Fiscal Year 2005

- Audit of the Reactor Program System

- Audit of NRC's Telecommunications Program

- Audit of NRC's Policy and Practices Concerning Camera Cell Phones

Investigation

- Theft of Cash and Credit Cards at NRC Headquarters

CHALLENGE 6
Administration of all aspects of financial management.

Sound financial management includes implementation of new internal control requirements, preparation of financial statements in accordance with applicable requirements, and efficient and effective procurement operations. A brief discussion of these areas follows.

New Internal Control Requirements

NRC's challenge is to take systematic and proactive measures to implement new OMB internal control requirements which become effective in FY 2006. The Office of the Chief Financial Officer serves as the agency lead to implement the new requirements of OMB Circular No. A-123 Revised, *Management's Responsibility for Internal Control*, December 21, 2004. This Circular provides Federal managers with guidance on improving the accountability and effectiveness of Federal programs and operations by establishing, assessing, correcting, and reporting on internal control. The Circular, among other things, emphasizes the need for integrated and coordinated internal control assessments that synchronize all internal control-related activities. It provides updated internal control standards, as well as new specific requirements for conducting management's assessment of the effectiveness of internal control over financial reporting.

Financial Statements

While NRC received an unqualified audit opinion on its FY 2004 financial statements, the audit opinion on the agency's FY 2003 financial statements was revised from unqualified to qualified due to the lack of evidential matter to support the completeness of accounts receivable and revenue. The agency's independent auditors characterized NRC's fee billing system as a material weakness.

Procurement

NRC's procurement of goods and services must be made in accordance with Federal regulations and with an aim to achieve the best value for the agency's dollars in a timely manner. Agency policy provides that the NRC's procurement of goods and services support the agency's mission and be planned, awarded, and administered efficiently and effectively. Among the numerous challenges facing the Agency in these areas are:

- Hiring and training new contract personnel,

- Keeping current with the Federal Acquisition Regulation (FAR),

- Ensuring adequate competition in awarding contracts,

- Considering past-performance in awarding contracts,

- Identifying the need for contract audit services, and

- Monitoring purchase card transactions.

An additional challenge facing the agency is the need to focus efforts on compliance with the FAR time standards for closing expired contracts and prompt deobligation of excess funds, which would make those funds available for other agency priorities.

Related Office of the Inspector General Work

Audits

- Results of the Audit of the U.S. Nuclear Regulatory Commission's Financial Statements for Fiscal Years 2004 and 2003

- Independent Auditors' Report on the U.S. Nuclear Regulatory Commission's Special-Purpose Financial Statements as of September 30, 2004, and for the Year then Ended

- Independent Accountants' Report on the Application of Agreed-upon Procedures for Federal Intragovernmental Activity and Balances as of September 30, 2004

- Review of NRC's Implementation of the Federal Managers' Financial Integrity Act for Fiscal Year 2004

- Audit of the Budget Formulation Process

- Audit of NRC's Contract Closeout Process

Investigations

- Fraudulent Workers' Compensation Program Claim by NRC Employee

- Fraudulent Workers' Compensation Program Claim of Former NRC Employee

- Failure to Report Billing Errors to Independent Auditors

- Mischarging of Costs by NRC Contractor

- False Reporting of Recyclable Material by Government

- Misuse of Government Travel Card by SBCR Employee

- NRC Mischarging EPRI for Reviewing Generic Products

- Misuse of NRC Travel Card by ACRS Member

- Misuse of NRC Travel Card by OIP Employee

- Fraudulent Use of NRC Purchase Card

- False Claim of Small Business Status by NRC Licensee

- Fraudulent Use of NRC Travel Card Account Number by Persons Unknown

- Misuse of NRC Citibank Travel Card by Region I Employee

CHALLENGE 7
Communication with external stakeholders throughout NRC regulatory activities.

The NRC believes that nuclear regulation is the public's business and, therefore, it should be transacted in an open and candid manner in order to maintain the public's confidence. Therefore, management should ensure that there are adequate ways of communicating with and obtaining information from external stakeholders that may have a significant impact on the agency achieving its goals. NRC established a strategic goal which ensures openness that expressly recognizes that the public must be informed about, and have a reasonable opportunity to participate in, the regulatory processes. Because of the nature of its business, the agency needs to interact with a diverse group of external stakeholders (e.g., the Congress, general public, other Federal agencies, and various industry and citizen groups) with clear, accurate, and timely information about NRC's regulatory activities. This continues to be a challenging task.

To this end, the agency has initiated efforts to improve cooperation and also enhance public outreach with specific stakeholders. NRC's Congressional Outreach program has produced favorable results. The program was initiated in 2005 and is directed at ensuring that the Congressional District offices are informed of NRC's activities in their

Districts. Information discussed in this venue is broad and can include such items as security, high-level waste, spent fuel storage, NRC organization, and programs such as reactor oversight, materials, and Agreement States.

Related Office of the Inspector General Work

Audits

- Audit of NRC's Baseline Inspection Program
- Audit of NRC's High-Level Waste Program
- Audit of NRC's Generic Communications Program
- Audit of the Reactor Program System

Investigations

- Inaccurate Information Provided by NRC to Congressman
- Mishandling of Allegation by NRR Staff

CHALLENGE 8

Intra-agency communication (up, down, and across organizational lines).

Effective intra-agency communications should occur with information flowing up, down, and across the organization. Information should be communicated to management and others within the organization who need it and in a form, and within a time frame, that enables them to carry out their responsibilities.

NRC has instituted various actions to improve its internal communications over the past year. The Director of Communications and technical communications assistants are working to continually improve this area. The agency continues to produce electronic "EDO Updates." These represent timely and succinct communications between the EDO and the entire staff. NRC's internal Web site addresses different types of employee concerns. NRC continues to hold "All Employees" meetings as a mechanism for direct two-way communication between the Commission and agency staff. Also, NRC's Strategic Plan stresses the importance of the role of internal communications in achieving the agency's mission and goals.

The Office of the Executive Director for Operations' (OEDO) internal web page provides various guidance to the staff, including guidance on communicating with the Commission, the OEDO mission and history, new and archived EDO updates, senior manager biographies, and OEDO staff contacts for various NRC program offices and topics.

The agency has made progress on the "roadmap initiative" which is intended to be a planning tool used to give the EDO's office a six month view of office and regional products, allowing early incorporation of senior management views, and reducing burden by minimizing re-writes and last minute changes in direction. To improve connectivity, Office Directors place their monthly roadmap reports in an ADAMS folder for access by other Office Directors and Regional Administrators to determine whether there are products that may require input or coordination.

Related Office of the Inspector General Work

Audits

- Audit of NRC's Baseline Inspection Program
- Audit of NRC's High-Level Waste Program
- Audit of the Budget Formulation Process

Investigation

- Failure to Report Billing Errors to Auditors

CHALLENGE 9
Managing human capital.

NRC faces current and emerging staffing challenges that could affect its ability to maintain the skills base needed to carry out the agency's mission. One of the challenges faced, along with the rest of the Federal Government, is an aging workforce. Retirement accounts for just over half of NRC's attrition, which most directly depletes the knowledge base. This makes identification of probable retirements and plans for successful replacement of those skilled individuals a high priority.

The challenge to be met by the NRC is preparing to replace an increasing number of individuals who become eligible to retire, taking with them valuable skills and institutional knowledge. However, not all individuals retire immediately upon eligibility. At NRC, on average, employees stay about four years beyond their retirement eligibility date.

NRC's workforce must possess detailed knowledge and specialized technical skills to fulfill its public health and safety mission. To maintain this expertise, NRC will need to build its human capital in the technical, financial, and administrative areas. In its Strategic Plan, NRC identified the management of human capital as a major challenge because of declining workforce numbers, loss of institutional knowledge and critical skills, and a shrinking labor pool.

NRC periodically assesses its human capital situation looking for ways to make improvements to support the achievement of its mission and goals. Agency efforts in critical skills staffing and training/development are described as follows:

Critical Skills Staffing

NRC currently uses a wide variety of human capital policies and programs for recruiting, hiring, training and development, and retention. The agency is challenged by preparing for growth in current and emerging work requirements including license renewals, applications for power uprates, potential licensing for high-level waste, and new reactor projects. These factors will require an increase in staff resource needs. Uncertainty of licensee schedules complicates the agency's efforts to have staff available with the right skills at the right time.

NRC measures the skills supply and demand using a strategic workforce planning web-based tool. The system provides a means for managers to specify their near-term and long-term skill needs and provides employees with a way to indicate their level of expertise in these skills areas. The agency completes an annual information call where managers identify continuing and newly anticipated skill gaps. Once this information is analyzed, the Office of Human Resources works with managers to plan and implement skill gap closure strategies.

Training/Development

The purpose of strategic planning for training and development is to ensure that processes, infrastructure, evaluation, and feedback methodologies are in place so that the agency's training and development activities mature and maintain the critical knowledge competencies needed to execute the agency's strategic mission. To accomplish this, the Office of Human Resources is developing a Strategic Plan for training and development, which outlines a training and development vision, mission, and strategic outcome for the agency. The focus is on alignment with agency goals and strategies and to provide for training support for staff. NRC's plan in this area outlines a number of goals, some of which include:

Focus on optimizing resources spent to conduct high quality training to meet the needs of a diverse workforce,

- Provide and support comprehensive, integrated, competency-based programs for staff, and

- Use training resources (expertise, facilities, equipment, and analytical tools) to effectively support other agency programs, including incident response and international activities.

Related Office of the Inspector General Work

Audit

- Audit of NRC's Baseline Inspection Program

CONCLUSION

Although the nine challenges identified in this report on the last page are distinct, they are also interdependent. One of the OIG's strategic goals is to improve the economy, efficiency, and effectiveness of NRC corporate management. The Inspector General's identification of the most serious management challenges facing the agency and the OIG's commitment to ensuring the integrity of NRC programs and operations help achieve this goal. Further, as evidenced by this review, the agency continues to take steps to address the management challenges through planning and in day-to-day operations.

Most Serious Management Challenges Facing the Nuclear Regulatory Commission as of September 30, 2005
(as identified by the Inspector General)

Challenge 1   Protection of nuclear material used for civilian purposes.

Challenge 2   Protection of information.

Challenge 3   Development and implementation of a risk-informed and performance-based regulatory approach.

Challenge 4   Ability to modify regulatory processes to meet a changing environment.

Challenge 5   Implementation of information resources.

Challenge 6   Administration of all aspects of financial management.

Challenge 7   Communication with external stakeholders throughout NRC regulatory activities.

Challenge 8   Intra-agency communication (up, down, and across organizational lines).

Challenge 9   Managing human capital.

ACTIONS TO ADDRESS THE NRC'S MANAGEMENT CHALLENGES

This appendix lists the management challenges that the NRC's Inspector General identified for FY 2005 in an October 4, 2004, letter to Chairman Diaz, and discusses the actions that the agency has taken to address those challenges.

1.   Protection of nuclear material used for civilian purposes

The NRC continues to enhance current security measures to ensure adequate protection of the Nation's nuclear materials and facilities. In FY 2005, the agency used a risk-informed approach to assess the potential vulnerabilities of civilian nuclear facilities and activities. The agency coordinated its activities with the Homeland Security Council, the Department of Homeland Security, the Federal Bureau of Investigation, the Department of Energy, the Defense Threat Reduction Agency, and other agencies.

The NRC's comprehensive oversight of the security and safeguards of NRC-licensed nuclear facilities and activities resulted in the following significant improvements in FY 2005:

The NRC issued a rule on fitness-for-duty and an order on access authorization. These actions, together with previously issued guidance on the revised design basis threat and Commission orders to implement security enhancements, will represent a significant advance in security planning.

The NRC revised its baseline inspection program for the physical protection cornerstone of the Reactor Oversight Process. This revised baseline inspection program reflects changes imposed by the Commission's orders in the areas of access authorization, fatigue, security officer training and qualification and the design basis threat. The NRC began implementing the revised baseline inspection program in FY 2004. The NRC also developed improved performance indicators and a revised Significance Determination Process to measure licensees' security performance more effectively.

Consistent with the Commission's orders revising the design basis threat, each licensee that operates a power reactor or a Category I fuel cycle facility (strategic special nuclear material in any combination in a quantity of 5000 grams or more computed by the formula grams = (grams contained U-235) + 2.5 (grams U-233 + grams plutonium) submitted a revision of its physical security plan or plans, contingency response plan or plans, and training and qualification plan or plans for NRC staff approval in April 2004.

The NRC staff finished reviewing these plans by October 2004. Full implementation of these plans significantly increased the licensees' ability to defend their facilities against a comprehensive set of adversary characteristics.

In FY 2004, the agency completed a series of assessments that provided the technical bases for additional mitigative measures that may be required to protect the Nation's nuclear materials and facilities.

In FY 2004, in collaboration with the Department of Homeland Security, the Department of Energy, and other Federal agencies, the NRC continued to assess the potential use of radioactive sources in radiological dispersion devices and to identify necessary enhancements in the control of radioactive sources. As a result, the agency has enhanced the security requirements for licensees who hold radioactive material(s) designated as radionuclides of concern. The NRC staff also worked with the Agreement States to develop appropriate enhancements for lower priority high-risk sources. In FY 2005, the agency moved to enhance its nuclear material management and safeguards system and codified a list of radioactive sources for import and export in 10 CFR Part 110. This list was developed to reflect efforts by the International Atomic Energy Agency and the Department of Energy/NRC working group. The NRC expects to move forward with its national source tracking system rulemaking in FY 2006. In addition, working with the Homeland Security Council, its oversight committees in Congress, the Administration, and other Federal agencies, the NRC continues to support legislative proposals to enhance the security of nuclear materials and facilities. A number of NRC's security-related legislative proposals (and others not requested by the NRC) were included in H.R.6, the Energy Policy Act of 2005, enacted on August 8, 2005. For example, Section 651(d) of the act establishes a new task force of Federal agencies, headed by the Chairman of the NRC, to evaluate and provide recommendations to the Congress and the President on security of radiation sources in the United States from potential terrorist threats. Section 651(d) also requires the NRC to enter into an arrangement with the National Academy of Sciences to conduct a study of industrial, research and commercial uses for radiation sources and to submit the results of the study to the Congress within 2 years of enactment of this provision. Section 652 expands the requirement to conduct fingerprinting, for criminal history record checks, to broader classes of entities and individuals. Section 653 authorizes NRC to allow security guards at certain licensed facilities to possess more powerful weapons and Section 654 makes it a Federal crime to introduce, without authorization, weapons or explosives into NRC-regulated facilities designated by the Commission.

The act also requires NRC to undertake several security-related activities which the agency has already done or is doing. Section 651(d) requires the agency to issue regulations governing the import and export of radioactive sources and also to issue regulations establishing a radioactive source tracking system. As noted above, NRC issued a final rule on export and import of radioactive sources and published a proposed rule on its National Source Tracking System in FY 2005. Section 651 (a) requires the NRC to conduct security evaluations, including force-on-force exercises, at least once every three years at facilities designated by the NRC and to report to Congress annually on the evaluations. While the reporting requirement to Congress is new, the Agency has been conducting force-on-force exercises for some time at reactors and Category 1 fuel facilities.

In FY 2005, the NRC staff developed a plan for implementation of these and other applicable requirements of the act in FY 2006 and beyond.

The NRC expanded and strengthened its information security program, which permits routine sharing of classified and sensitive unclassified information with authorized representatives up to the "secret national security information" level. The NRC has significantly enhanced secure communication capabilities at headquarters and the regional offices. In so doing, the NRC ensured timely communication among authorized individuals while effectively protecting classified and sensitive unclassified information (both internally and externally) through the use of administrative procedures and requirements that are consistent with Federal law and national programs. In FY 2005, the NRC finished installing, certifying, and accrediting 12 secure video sites in headquarters and the regions. The NRC continued to interact, communicate, and coordinate with other Federal, State, and local agencies, and the international community with respect to homeland security, emergency response, and integrated response planning. The NRC successfully responded to several unique events, as well as two inadvertent overflights of the Washington, D.C. metro area. The NRC continues to work with the Department of Homeland Security and other Federal agencies to implement and administer a National Incident Management System and a unified National Response Plan in accordance with Homeland Security Presidential Directive 5, "Management of Domestic Incidents." The NRC continued to implement upgrades to the agency's Incident Response Operations Center in FY 2005 with a top to bottom review of the entire incident response center and plans, and the development of a long term Incident Response Improvement Plan, which will significantly enhance the agency's response center. The NRC established an alternative incident response center at one of the agency's regional offices. This alternative center has all of the capabilities of the headquarters operations center, in the event of a loss of the headquarters facility.

In FY 2004, the NRC completed a pilot force-on-force exercise program which administers an exercise to provide a more realistic test of plant capabilities to defend against an adversary force. This effort reduced artificialities and increased the realism of the exercises. The agency has also used the results of the expanded pilot exercises to revise the staff's exercise program and improve the NRC's processes for assessing licensees' readiness to respond to the design-basis threat. The exercise provides details regarding specific adversary characteristics against which security forces at nuclear power reactors and Category I fuel cycle facilities (facilities that process strategic special nuclear material) need to be protected.

The NRC met regularly with industry representatives to catalog and discuss the lessons learned from the exercises, documenting the staff's and the industry's perspectives. In implementing the force-on-force program, the NRC increased the frequency of force-on-force drills at power reactor facilities from once every eight years to once every three years. As intended, force-on-force exercises have been a primary means to conduct performance-based testing of a licensee's security plan and its ability to prevent radiological sabotage. In FY 2005, the agency completed a transitional force-on-force exercise program and followed up with full program implementation. In FY 2005, the exercise evaluated one-third of its licensees through the force-on-force exercise program.

Finally, the agency completed its interactions with the National Academy of Sciences on a highly visible spent fuel storage report and reported the results to the Congress.

2. Protection of information

In FY 2005, the NRC faces several challenges to ensure compliance with the Federal Information Security Management Act. The NRC's current certification process is revealing previously undiscovered security risks and some slippage in system certification. The NRC staff has increased its efforts to provide more rigorous independent review, testing, and evaluation of major system security plans. The staff is also developing standardized information technology security solutions to minimize costs and provide the right level of protection, while complying with the Federal Information Security Management Act. The NRC has an effective information technology security training and awareness program. All employees are required to complete an online information technology security training course, and NRC information systems security officers and other employees and support contractors with significant security responsibilities are required to complete a more advanced online technical security course. The NRC maintains an information technology security Web page to promptly inform NRC employees of information technology security issues. The NRC has a robust incident reporting program in place, and files monthly reports to the Federal Computer Incident Response Center.

New policies for Federal agency public Websites: On December 17, 2004, the Office of Management and Budget issued M-05-04, "Policies for Federal Agency Public Websites," to heads of executive departments and agencies. The document contains new e-Government guidance to bring agencies' public web sites into compliance with Federal information resource management laws and policy. These are new requirements to ensure that agencies manage their Federal agency public Websites as part of their information resource management program. The guidance also identified several new Office of Management and Budget policies with which agencies must be fully compliant by December 31, 2005. Similarly, on January 21, 2005, the National Archives and Records Administration issued to Federal Agency Contacts, Modern Records Programs 14.2005, "NARA Guidance on Managing Web Records." The National Archives and Records Administration guidance described agency Website operations as an integral part of an agency's program that contains both content and administrative Federal records that agencies must manage properly. To implement the new Office of Management and Budget and the National Archives and Records Administration guidance, the NRC has developed an action plan to evaluate its Web records, conduct risk assessments to identify improperly managed records, and prepare records disposition schedules for the content and administrative Web records, as appropriate.

3. Development and implementation of a risk-informed and performance-based regulatory oversight approach

For many years, the NRC has developed and adapted methods for undertaking probabilistic risk assessments and performance assessments to enable the agency to better understand the risks of licensed activities. During FY 2005, the NRC used these methods by supporting the development of calculation tools and experimental results to provide the basis for risk-informed regulation. Risk-informed regulation uses risk analysis, along with engineering studies, to focus regulatory and licensee attention on design and operational issues in a manner that is commensurate with the risks that the issues pose to public health and safety. Incorporating risk analysis into regulatory decisions improves the regulatory process by focusing NRC and licensee attention and activities on the areas of highest risk, thereby reducing unnecessary burden on licensees and increasing efficiency and effectiveness in the use of agency resources.

The NRC's FY 2004–FY 2009 Strategic Plan states that future challenges to the agency's regulatory climate are expected to require adjustment to both internal and external factors, such as the use of risk-informed and, as appropriate, performance-based regulations. To further the goal of broadly applying risk techniques to the agency's regulatory processes, the NRC developed the Risk-Informed Regulatory Implementation plan. The NRC has included milestones in the Risk-Informed Regulatory Implementation plan as performance measures in working toward making the agency's activities and decisions more effective, efficient, and realistic. During FY 2005, the NRC has taken agencywide actions across the agency to meet this challenge, as described in the following paragraphs.

Nuclear Reactor Safety

Based on its assessment of stakeholder feedback and the results and lessons learned from self-assessments during FY 2005, the NRC staff believes that the Reactor Oversight Process has satisfied the Commission's direction to develop an oversight process that is more objective, risk-informed, understandable, and predictable than previous processes. The staff plans to continue annual self-assessments and report on lessons learned from the implementation of the Reactor Oversight Process to the Commission. During FY 2005, the staff also revised inspection procedures to incorporate recommendations from the Davis-Besse Lessons Learned Task Force and tested the effectiveness of a new procedure for engineering design inspections that focuses on aspects of the plant design that represent a relatively high degree of risk and for which there appear to be relatively low margin. The procedure was implemented at one site in each of the four NRC regions. The staff concluded that aspects of the pilot inspection approach resulted in improvements that should be incorporated into the baseline inspection program. The staff plans on incorporating these attributes into a revised baseline inspection procedure to be implemented beginning January 2006.

Development of the Mitigating Systems Performance Index continued during FY 2005. These performance indicators provide a more accurate indication of the risks associated with changes in the availability and reliability of important safety systems. The index is based on risk-significant functions and uses plant-specific risk models and importance measures. The staff has completed a one year pilot of the Mitigating Systems Performance Index. In SECY-04-0053, the staff documented several technical issues that were unresolved at the completion of the pilot program. Those issues have now been resolved and the staff agreed to move forward with Mitigating Systems Performance Index Implementation. The staff and industry are working together to address implementation issues. A current target date for full implementation is set for January 2006.

The staff continues to work on initiatives, defined by the Significant Determination Process Task Action Improvement Plan, to address timeliness and other improvements to the Significance Determination Process. The Significance Determination Process is used to assess the safety significance of reactor events and inspection findings.

In FY 2005 the NRC staff added a new methodology to the Significant Determination Process which provides NRC inspectors the tools needed to assess the risk significance of identified fire protection issues and provides the tools needed to assess the risk significance of inspection findings related to licensee assessment and management of risk associated with performing maintenance activities under all plant conditions. Additionally, a significance determination process to assess inspection findings related to spent fuel storage is in development and the need for a new methodology is being examined for assessing findings associated with the performance of the on-site fire brigade.

During FY 2005, consistent with the Commission's policy statements on technical specifications and the use of probabilistic risk assessment, the NRC and the industry continued to develop risk-informed improvements to the current system of technical specifications. These improvements are intended to maintain or improve safety while reducing unnecessary burden and to bring technical specification requirements into congruence with the Commission's other risk-informed regulatory activities. In FY 2005, the NRC approved a variety of risk management technical specification initiatives, including: (1) allowances for a risk-informed evaluation to determine whether it is preferable to shut down or to continue to operating a reactor plant under certain degraded conditions, and (2) flexibility in determining the required actions to be taken when certain support equipment is not operable but can still function, (3) flexibility in determining short term technical specification required actions end states for repairing inoperable equipment, and, (4) flexibility in transitioning up in mode to power operation with inoperable equipment that is to be restored to operable status within the technical specification required action completion time. The NRC continued the review of an industry proposal for risk management of allowed outage times for technical specification equipment.

In February 2004, the NRC has issued for trial use Regulatory Guide 1.200, "An Approach for Determining the Technical Adequacy of PRA Results for Risk-Informed Activities," and the associated Standard Review Plan Chapter 19.1, "Determining the Technical Adequacy of Probabilistic Risk Assessment Results for Risk-Informed Activities." This Regulatory Guide provides guidance to licensees concerning the quality needed for probabilistic risk assessment information used in risk-informed activities. A trial use period was scheduled to test the implementation of the guide through a variety of different risk-informed applications. Five licensees volunteered to participate as a pilot plant during the period of trial use. The pilot applications conducted for the trial use period were completed in March 2005. In July 2005, American Society of Mechanical Engineers issued an addendum to its standard based on lessons learned from the pilots. Pending the timely issuance of American Society of Mechanical Engineers' revised standard, RG 1.200, Revision 1, is scheduled to be issued for use in FY 2006.

In FY 2005, the NRC staff completed the development of the technical basis necessary to support a risk-informed rulemaking effort to modify the pressurized thermal shock screening criteria in 10 CFR 50.61. The reports which document this technical basis will be published in December 2005 and represent the final revision of the draft technical basis report which was issued on December 31, 2002. This technical basis was reviewed at various stages by the NRC's external stakeholders, a select external peer review panel of technical and regulatory experts, the Advisory Committee for Regulatory Safeguards, and NRC technical staff. At this time, the staff expects to recommend that the rulemaking process to revise 10 CFR Part 50.61 be initiated and to submit a rulemaking plan to the Commission in FY 2006.

Staff efforts to develop an alternative risk-informed and performance based fire protection option for nuclear power plants carried forward into FY 2005. An industry standard, National Fire Protection Association 805, "Performance-Based Standard for Fire Protection for Light-Water Reactor Electric Generating Plants, 2001 Edition," was issued in April 2001. The final rule to incorporate National Fire Protection Association 805 in 10 CFR Part 50.48) was published in the *Federal Register* in June 2004. The staff worked with the industry to complete development of the implementation guidance for National Fire Protection Association 805 that was endorsed by the NRC via a regulatory guide. The regulatory guide was published in the *Federal Register* for a 60-day public comment period in October 2004. The staff plans to issue the final regulatory guide in March 2006. The industry is preparing a revision to the implementation guide to incorporate additional NRC guidance. NRC staff are also developing accordant inspection procedures that are expected to be completed in December 2005.

Efforts persist to resolve issues related to the requests for additional information on the Risk Management Technical Specifications Initiative 4b, Risk Informed Completion Time (including the Industry Risk Management Guide, the Combustion Engineering pilot proposal, Technical Specifications Task Force No. 424, and the South Texas Project pilot submittal). The industry provided a draft Industry Risk Management Guide and the Combustion Engineering Owner's Group single system pilot proposal, Technical Specifications Task Force-424, on January 21, 2003. In addition, the South Texas Project submitted a whole-plant proposal in support of the Risk Management Technical

Specifications Initiative 4b and on Probabilistic Risk Assessment quality, Regulatory Guide 1.200, in August 2004. The NRC staff has issued requests for additional information for the Industry Risk Management Guide, the Combustion Engineering pilot proposal, Technical Specifications Task Force-424 and South Texas Project submittals, and briefed the Advisory Committee Reactor Safeguards full committee in May 2004 and the Advisory Committee Reactor Safeguards subcommittee in June 2005. The expected completion date for CE initiative 4b, involving single High Pressure Safety Injection system application, is summer 2006.

The NRC staff continued its work to improve the requirements contained in 10 CFR 50.46 as they relate to analysis of design-basis large-break loss-of-coolant accidents and associated emergency core cooling performance and analysis. The NRC staff also proceeded with a number of related activities, including developing frequency estimates for loss-of-coolant accidents and working on a proposed rule to allow use of an alternative maximum break size. The development of a risk-informed approach to 10 CFR 50.46 has the potential to improve significantly the effectiveness of regulatory oversight related to emergency core cooling system performance. In August, 2004, the staff published a conceptual basis and draft language for the proposed rule and held a public meeting. During FY 2005, the staff evaluated information received at the public meeting and provided the Commission with a memorandum summarizing the proposed rule and providing draft language on October 22, 2004. The staff also met with the Advisory Committee Reactor Safeguards subcommittee once and with the full committee three times. An Advisory Committee Reactor Safeguards letter dated March 14, 2005 was supportive of the proposed rule. The proposed rule making package went to the Commission on March 29, 2005. The Commission issued a Staff Requirement Memorandum on July 29, 2005, approving the rule with changes. The rule went to administrative publications in mid October 2005.

With regard to achieving coherence among risk-informed activities, the staff issued a draft revision to the Coherence Plan in December 2004 for internal review and comment. However, in June 2005, further work on development of a revised coherence plan was suspended. The coherence plan has been subsumed by staff efforts to respond to recent Commission direction contained in Staff Requirement Memorandum 050405 to develop a formal program plan to make a risk-informed and performance-based revision to 10 CFR Part 50. The staff plans to inform the Commission of possible approaches to developing this formal program plan in FY 2006.

In 2001, the NRC staff initiated a program with the objective of creating an environment in which risk-informed methods are integrated into staff activities, and staff plans and actions are naturally based on the principles of risk-informed regulation. The program has four phases: (1) evaluate the current environment, (2) design an improved risk-informed environment, (3) implement changes to achieve the target environment, and (4) assess effectiveness of environmental changes. Phases 1 and 2 have been completed. A plan for implementing changes to enhance the current environment was developed. Phases 3 and 4 are on hold, pending higher priority work (associated with Nuclear Security and Incident Response).

Nuclear Materials and Waste Safety

Over the past year, the NRC made significant progress toward increasing the use of risk insights and information where feasible and beneficial. The agency is currently developing guidance documents and risk guidelines to facilitate consistent and effective application of the risk-informed approach.

In FY 2004, the NRC completed feasibility/scoping study to identify human reliability analysis development needs for the wide range of situations encountered and activities performed by licensees subject to the Nuclear Materials Safety program. Based on this study, in FY 2005 the NRC has begun to prioritize human reliability analysis needs in the Nuclear Materials and Waste Safety program. In addition, the staff has begun developing human reliability tools and information to address a high priority need in the area of nuclear medical devices. Also, tasks have been initiated to develop human reliability tools and information to address a high priority need in the area of spent fuel handling.

The NRC identified nuclear material safety and safeguards regulatory applications that are amenable to increased use of risk insights and evaluated recommendations to improve the effectiveness and efficiency of the byproduct materials program. Several byproduct materials guidance documents were revised to incorporate risk insights, specifically those addressing technical assistance requests, Consolidated Guidance about Materials Licensees (NUREG-1550), and Inspection Manual Chapter 2800, "Materials Inspection Program." In FY 2005, the NRC continues its efforts to risk-inform the nuclear materials program. The NRC revised one guidance document, NUREG-1556, Volume 9, to incorporate risk insights to conform to the amended training and experience requirements for medical use of byproduct material.

In addition, the NRC has been working during FY 2005 to revise the Fuel Cycle Oversight Program in accordance with new 10 CFR Part 70 risk-informed regulatory requirements. The NRC is making progress in developing and implementing methods for risk-informed licensing reviews and risk-informed inspections.

The NRC continues to incorporate lessons learned into guidance development so the agency can apply risk-informed approaches consistently and effectively where appropriate. In FY 2005, the staff drafted a guidance document on Risk-Informed Decision Making for Materials and Waste Applications, to aid in using a risk-informed decision-making process on applicable regulatory issues in the Nuclear Materials Safety area. This document is currently undergoing final reviews prior to issuance.

In FY 2005, the NRC continued work on a probabilistic risk assessment of a dry cask storage system. This probabilistic risk assessment study provides a method for quantifying the risks of dry cask storage of spent nuclear fuel and provides insights for improved decision-making concerning regulatory activities associated with 10 CFR Part 72.

In FY 2005, the NRC staff issued Revision 1 of NUREG-1762, "Integrated Issue Resolution Status Report." This report consolidates information on the closure of issues concerning the prospective license application for a geologic repository at Yucca Mountain. The NRC and the Center for Nuclear Waste Regulatory Analysis completed the risk analyses for risk insights. The analyses enhanced understanding of significant issues in the risk insights baseline. The staff concluded that no modifications of the April 2004 Risk Insights Baseline Report were required. The Risk Insights Baseline Report supported the completion of pre-licensing issue resolution agreements and is being used by the staff in preparing for the review of a potential license application for the Yucca Mountain high level waste repository.

In FY 2005, the NRC staff continued regulatory improvements to resolve the issues that were identified in the staff's 2003 evaluation of implementation of 10 CFR Subpart E, the license termination rule. These improvements better incorporate risk insights in implementing the License Termination Rule. The staff has begun the process for developing regulations to prevent future legacy sites and is revising the decommissioning guidance for the following issues: restricted use/institutional controls; on-site disposal approvals; more realistic exposure scenarios; and the use of intentional mixing of soil. The staff conducted a decommissioning workshop to seek early licensee and other stakeholder input on the scope of this guidance and plans on publishing draft guidance in September 2005 for public comment. The staff has also implemented a communication strategy to expand stakeholder involvement by conducting the decommissioning workshop noted above, establishing a State working group for the guidance development, publishing a decommissioning program brochure, and enhancing the decommissioning Web page. The staff is also exploring new ways to share decommissioning experience and lessons learned with other groups involved with decommissioning such as Agreement States, the Department of Energy, and industry groups.

4. Ability to modify regulatory processes to meet changing external demands

The NRC uses its planning, budgeting, and performance management process to integrate the agency's regulatory processes and ensure that the agency is able to respond to changes in its environment. Each year, the Program Review Committee holds planning sessions to ensure that the Commission's regulatory processes are integrated and resources allocated where needed. The Commission approves the Program Review Committee's plans during the budget process. In addition, the Executive Director for Operations holds meetings to ensure agencywide integration.

The NRC issues regulations considered necessary to ensure that licensees operate their reactor facilities in a safe manner and that the agency meets its strategic goal to protect the public health and safety. Any rule imposing requirements needs a backfit analysis (in accordance with the backfit rule set forth in 10 CFR 50.109) to demonstrate that the requirements either are necessary for adequate protection or are cost-beneficial safety enhancements. The completed regulatory actions reflecting this position during FY 2004 included the rulemaking on performance-based risk-informed fire protection and the risk-informed 50.44 Rulemaking.

Quarterly meetings of the Probabilistic Risk Assessment Steering Committee ensure that risk-informed activities are integrated agencywide. Similarly, the agencywide participation of NRC managers on the Research Effectiveness Review Board ensures that the agency's research program effectively meets agencywide needs.

The NRC's Risk Steering Committee provides guidance and sets expectations for implementing risk-informed initiatives in the Nuclear Materials and Waste Safety programs. This committee is comprised of agency experts who offer guidance in risk-informing initiatives. These experts also provide peer review of risk-informed products.

The NRC and representatives of the Nuclear Energy Institute hold periodic Fire Protection Issues Management Meetings. These meetings provide a forum through which the NRC and the industry can identify and prioritize emerging fire protection issue and develop resolutions.

The NRC's Rulemaking Coordinating Committee, established in 1998, ensures that the agency's rulemaking process is consistent agencywide. The primary focus of the Rulemaking Coordinating Committee is to ensure consistency in the methods used to develop and promulgate rules and to facilitate initiatives for improving the rulemaking process.

The staff continued to prepare for receipt of the Department of Energy's anticipated high-level waste repository license application and the associated hearings. This cooperative effort involves putting the systems and processes in place needed to meet the 3-year deadline.

5.   Implementation of information resources

The NRC's actions to address this management challenge in FY 2005 are discussed in detail in the section of Chapter 2 on the President's Management Agenda. Please read the discussion of Expanded Electronic Government under the Federal Information Security Management Act.

6.   Administration of all aspects of financial management

The NRC's actions to address this management challenge in FY 2005 are discussed in detail in the section of Chapter 2 on the President's Management Agenda. Please read the discussion of Improved Financial Management.

7.   Communication with external stakeholders throughout NRC regulatory activities

The NRC issued guidelines for effectively communicating risk-related information to external stakeholders ("Effective Risk Communications," NUREG/BR-0308, dated January 2004). The document provides easy-to-use guidance for agency staff and management on NRC-specific communication topics and situations that deal with risk to ensure the agency's openness with the public. The guidance contains practical suggestions, tailored to the NRC's needs, that reflect the risk communication best practices learned from researchers, trainers, and practitioners from numerous Federal, State, private, and educational organizations.

Nuclear Reactor Safety

The NRC developed and implemented an array of plans for communications on the recent events at the Davis-Besse Nuclear Power Station and the Vermont Yankee Generating Station, fire protection, and Generic Safety Issue 191, "Assessment of Debris Accumulation on Pressurized-Water Reactor Sump Performance."

In FY 2005, the NRC's license renewal program staff conducted 26 public meetings on the NRC's license renewal application review process and environmental issues. As of June 2005, the NRC had conducted an additional 17 license renewal public meetings. These meetings afforded the NRC the opportunity to solicit stakeholder comments. The meeting also allowed a meaningful exchange of information with external stakeholders on the safety and environmental effects of continued operation, the license renewal process, and opportunities for public involvement. The NRC held these meetings in the vicinity of those affected by the actions to be discussed.

In FY 2005, the NRC staff has held five public outreach meetings on the reactor vessel head degradation at the Davis-Besse Nuclear Power Station and the NRC's related response and evaluation. These meetings informed external stakeholders about the NRC's oversight activities and the Davis-Besse restart activities, and also gave citizens the opportunity to comment and ask questions.

The NRC also held a public meeting near the Vermont Yankee Nuclear Generating Station to discuss the NRC's power uprate review process and to obtain feedback from the public. The NRC also held public meetings near each nuclear power plant during FY 2004 to discuss the NRC's annual assessment of each plant's safety performance. These meetings gave external stakeholders information on each plant's safety performance and the NRC's role in ensuring safe operation.

In September and October 2003, the NRC received three early site permit applications for the Clinton, North Anna, and Grand Gulf sites. In FY 2004, the NRC held three scheduled public meetings to inform the respective communities of the NRC's regulatory role and the process for evaluating early site permit applications. In FY 2005, the NRC staff issued the draft safety evaluation report and draft environmental impact statement for all three applications. In June 2005, the staff issued the final safety evaluation report for the North Anna site, and the staff is scheduled to issue the two final safety evaluation reports in FY 2006. The staff is reviewing comments on the three draft environmental impact statements and is scheduled to issue the three final environmental impact statements in FY 2006.

Nuclear Materials and Waste Safety

During FY 2005, the NRC coordinated with the Department of Energy on several projects including the mixed-oxide fuel fabrication facility, and issues related to gas centrifuge uranium enrichment (e.g., the development of a memorandum of understanding on oversight of the U.S. Enrichment Corporation, Inc.) gas centrifuge facility.

During FY 2005, the NRC's Fuel Cycle Facilities Licensing and Inspection program staff has conducted 12 public meetings on significant regulatory issues. These meetings allowed the NRC to solicit stakeholder viewpoints and stakeholders to exchange information on issues such as the gas centrifuge facility licensing initiative and licensee performance reviews. Most of these meetings took place near the people affected by the actions.

In March 2005, the NRC issued a final rule to amend the agency's requirements for training and experience, as set forth in 10 CFR Part 35, "Medical Use of Byproduct Material." The staff also revised the guidance document "NUREG-1556, Volume 9, Consolidated Guidance About Materials Licenses, Program-Specific Guidance About Medical Use Licenses," to conform to the amended training and experience requirements. The NRC staff developed the rule with input from professional speciality boards and other members of the public and from the NRC's Advisory Committee on the Medical Uses of Isotopes. The staff also worked closely with the States to ensure a cooperative dialogue about the regulation of radioactive material. In addition, the NRC staff participated in the NRC's Advisory Committee on the Medical Uses of Isotopes meetings in October 2004 and April 2005 and the Conference of Radiation Control Program Directors meeting in April 2005.

The NRC staff has a generic communication plan for rulemakings. The primary goal of this plan is to ensure that the NRC conveys a consistent message to all internal and external stakeholders. The NRC also maintains a public Web site to communicate with stakeholders. This site provides links to pertinent documents, updates on current activities, and information on opportunities for stakeholder input.

During FY 2005, NRC continued its public outreach efforts for the proposed high-level waste repository at Yucca Mountain. NRC representatives also provided an overview of the agency's role in the potential licensing of the repository at several public outreach meetings in Nevada. The staff held a meeting with members of the public in Pahrump, Nevada, to discuss the Yucca Mountain project and a presentation to the National Conference of State Legislatures High-Level Waste Working Group.

Through May 2005, the NRC participated in 25 workshops, conferences, and town hall meetings with representatives of various Federal, State, and local agencies, international bodies, the nuclear industry; and public interest groups focused on spent fuel storage and transportation issues. The NRC conducted public meetings to seek input to inform national positions prior to significant meetings of the International Atomic Energy Agency concerning international transport regulations. The NRC updated and continued to implement the communications plan for spent fuel transportation, which provides a focused approach for public outreach and communication. The Spent Fuel Storage and Transportation program conducted a Licensing Process Workshop in February 2005 to: (1) introduce revised guidance for interaction with 10 CFR Part 71 and Part 72 applications ("rules of engagement") (2) discuss lessons learned and experience based on past practices (3) revise the interim staff guidance process to solicit stakeholder, public, and industry comments; and (4) solicit feedback and suggestions from applicants, stakeholders, industry, members of the press and other media, and members of the public on licensing process improvements. The Workshop addressed the agency's strategic goals of Openness, and Effectiveness in the agency's regulatory processes.

During FY 2005, NRC staff continued to participate in meetings of all five regional State groups on radioactive material transportation issues (i.e., the Southern States Energy Board, the Council of Governments Northeast High-Level Radioactive Waste Transportation Task Force, the Midwestern Radioactive Materials Transportation Committee, the Western Interstate Energy Board, and the Western Governors Association). The staff also participated in meetings held by the National Conference of State Legislatures' High-Level Radioactive Waste Working Group. NRC participation at these meetings helped to inform State public health and safety agency officials, as well as interested State legislative representatives. The meetings also allowed NRC staff to provide information to State-level representatives to address constituent safety concerns on transportation, to inform local and state transportation decisions, and served as an useful conduit for States to express concerns and seek information from the NRC.

The NRC conducted several public meetings with interested stakeholders on various sites or projects undergoing environmental review or scoping processes, including controlling the disposition of solid materials.

The NRC held numerous technical meetings with licensees to discuss issues associated with the decommissioning of their sites. These meetings were noticed in accordance with NRC requirements and guidance and were open to observation by members of the public. Meetings were held for the Westinghouse-Hematite, FMRI, Inc., Kerr-McGee-Cushing Cimarron, and Mallinkrodt material sites and for the Maine Yankee, Humbolt Bay, and Rancho Seco power reactor sites. In addition, NRC revised and enhanced the NRC's decommissioning Web page, published a brochure on the decommissioning process, and published the annual decommissioning report as a NUREG report.

The NRC maintains a public Web site to communicate with stakeholders. This site provides various links to pertinent documents, updates on current activities, and information on opportunities for stakeholder input. In FY 2005, the staff continued to add to the site semiannual reports on the rulemaking for controlling the disposition of solid material.

8.   Intragency communication (up, down, and across organizational lines)

The NRC staff routinely prepares communications plans on important topics. NRC prepares talking points and briefing papers on major activities to ensure consistency in key messages. The agency has also emphasized efficient meeting policies, promoted team building, and supported intra-office efforts to share important information agencywide.

The NRC continues to update and improve methods for meeting employees' information needs. Announcements have been streamlined. The offices will continue developing and updating their own individual Web pages linked to the agency's internal home page.

The NRC has established a Communications Council to coordinate and implement the agency's internal communication strategies and share best practices.

Internal NRC communications have increased, and a growing number of offices periodically issue internal electronic newsletters. The agency issues "EDO updates" in which the Executive Director for Operations regularly communicates important information to all agency staff, and the agency frequently issues memoranda to the staff on a variety of subjects. In addition, several individual offices have undergone detailed internal communication studies. These activities have included administering surveys, holding focus groups, and creating methods for collecting internal feedback.

The NRC regularly emphasizes good communication practices by agency managers. These practices include face-to-face communications, frequent feedback, and two-way communication. New leadership courses will also emphasize these practices and stress coaching and team building. In addition, many offices have created their own communication-related positions or teams address stakeholder concerns, promote good internal and external communication practices, and to address policy matters.

The NRC is reinforcing the agency's safety mission through e-mail messages, messages on the internal Web page, posters, memoranda, and other media. The agency is asking managers to emphasize that all performance measures support the safety goal because they allow the NRC and its licensees to focus on activities that are most important to safety. Agency managers will reinforce the linkage between the NRC's daily activities and the agency's safety mission.

In FY 2005, the offices involved in the Nuclear Reactor Safety program met periodically with intra-agency stakeholders to enhance communication and support functions. Offices responsible for this program also identified internal stakeholders as a target audience in their communications plans.

In addition, the Office of Nuclear Reactor Regulation developed a communications program for nuclear reactor regulation to support achievement of the agency's mission by providing tools, processes, and guidance to improve internal and external communication. The office has proposed and developed new and expanded communication efforts to encourage internal sharing of ideas, improve the flow of information to staff and management, and improve the timeliness, accuracy, and clarity of internal and external communications.

During FY 2005, the NRC continued to improve the communications among offices through periodic meetings. Communication between headquarters and regional offices continued to improve as a result of frequent conference calls at the staff and senior management levels, semiannual headquarters/regions counterpart meetings, trips, weekly informational e-mail messages, and the internal Web pages. The offices also continued to rotate staff and management assignments throughout the organization as a team building measure.

Communication between the Office of Nuclear Reactor Regulation and the agency's other program and support offices is improving as a result of the office's comprehensive communications program and agencywide support for the monthly Communication Council meetings. These meetings encourage sharing of best practices and lessons learned that apply agencywide. In addition, routine interactions with the NRC's Director of Communications assist the staff in defining, implementing, and continually improving communications.

The communication-related efforts undertaken by the Office of Nuclear Materials Safety and Safeguards include continuation of an active program for inter- and intraoffice rotational assignments, semiannual meetings with the Office of Nuclear Regulatory Research to review the status of ongoing research projects, monthly Decommissioning Management Board meetings to discuss decommissioning program activities, semiannual headquarters/regional counterpart meetings to discuss programmatic and technical issues in a focused, structured manner, and biweekly conference calls with the Regions and other internal stakeholders.

In February 2005, the Office of Nuclear Materials Safety and Safeguards conducted an independent spent fuel storage installation inspector counterpart meeting with inspectors and managers from headquarters and all the NRC regional offices. This meeting gave these NRC inspectors an opportunity to share experience, and discuss programmatic issues and challenges.

9. Managing Human Capital

The NRC's actions to address this management challenge in FY 2005 are discussed in detail in the section of Chapter 2 on to the *President's Management Agenda*. See the discussion on "Strategic Management of Human Capital."

MANAGEMENT DECISIONS AND FINAL ACTIONS ON OIG AUDIT RECOMMENDATIONS

The agency has established and continues to maintain an excellent record in resolving and implementing audit recommendations presented in Office of the Inspector General reports. Section 5(b) of the Inspector General Act of 1978, as amended, requires agencies to report on final actions taken on Office of the Inspector General audit recommendations. The following table gives the dollar value of disallowed costs determined through contract audits conducted by the Defense Contract Audit Agency and NRC's Office of the Inspector General. Because of the sensitivity of contractual negotiations, details of these contract audits are not furnished as part of this report. As of September 30, 2005, there were no outstanding audits recommending that funds be put to better use.

MANAGEMENT REPORT ON OFFICE OF THE INSPECTOR GENERAL AUDITS WITH DISALLOWED COSTS
For the period October 1, 2004-September 30, 2005

Category	No. of Audit Reports	Questioned Costs	Unsupported Costs
1. Audit reports with management decisions on which final action had not been taken at the beginning of this reporting period.	0	0	0
2. Audit reports on which management decisions were made during this period.	2	$42,079	0
3. Audit reports on which final action was taken during this report period.			
(i) Disallowed costs that were recovered by management through collection, offset, property in lieu of cash, or otherwise.	2	$42,079	0
(ii) Disallowed costs that were written off by management.	0	0	0
4. Reports for which no final action had been taken by the end of the reporting period.	0	0	0

MANAGEMENT DECISIONS NOT IMPLEMENTED WITHIN ONE YEAR

Management decisions were made before October 1, 2004, for the Office of the Inspector General audit reports discussed in the following paragraphs. As of September 30, 2005, NRC did not take final action on some issues.

NRC's License Fee Development Process Needs Improvement (OIG/99A-01)

December 14, 1999

The Office of the Inspector General recommended that the methodology for calculating the hourly rates for license fees be reevaluated to include the full-cost concept as embodied in Office of Management and Budget Circular No. A-25, User Charges, and Statement of Federal Financial Accounting Standards No. 4, Managerial Cost Accounting Standards, and that actual cost data are used to refine future rate calculations. The NRC implemented a cost accounting system in FY 2002, and cost data from this system was used as input to review the existing full-cost rate, including identification and assignment of direct and allocated indirect costs. In November 2003, NRC obtained contractor assistance to provide recommendations for improving NRC's license fee development process, including through the use of actual cost data to refine hourly rate calculations. Based on analysis of the contractor's recommendations, NRC developed procedures to calculate 10 CFR Part 170 hourly rates using actual cost data from the cost accounting system. The procedures have been piloted by using FY 2004 cost data to develop FY 2005 hourly rates. The NRC plans to use the procedures to calculate hourly rates each year, and to use those rates to recover the costs of activities under 10 CFR Part 170, beginning with the FY 2006 fee rule.

Review of Audit Follow-up System (OIG-00-A-05)

August 14, 2000

The Office of the Inspector General recommended that NRC revise the Management Directive and Handbook 6.1, *Resolution and Follow-up of Audit Recommendations*, governing resolution and follow-up of audits to reflect periodic scheduling standards for conducting analyses of audit recommendations to determine possible trends and system-wide problems and solutions, as required by Office of Management and Budget Circular A-50.

In addition, the Office of the Inspector General recommended that NRC assess its scheduling requirements for conducting audit follow-up reviews with the objective of conducting them on a consistent frequency. The NRC developed revisions of the management directive and handbook to incorporate these recommendations, as well as other changes, and the revisions are pending the Chairman's approval for issuance. The management directive and handbook were also revised to address the recommendations of OIG-00-E-09 as discussed in the paragraph in this section entitled "Special Evaluation of the Role and Structure of NRC's Executive Council (OIG-00-E-09)." Issuance of the revised management directive and handbook, which is expected by early 2006, will complete agency actions on the Office of the Inspector General's recommendations from this audit.

Special Evaluation of the Role and Structure of NRC's Executive Council (OIG-00-E-09)

August 31, 2000

The Office of the Inspector General recommended that NRC's management directives and communication mechanisms be updated to reflect the responsibilities and alignment of the Executive Director for Operations, the Chief Financial Officer, and the Chief Information Officer after the Commission decided on a management strategy for NRC's Executive Council. In January 2001, the Commission announced the abolishment of the Executive Council, although the Executive Director for Operations, Chief Financial Officer, and Chief Information Officer continue to meet periodically to ensure necessary communications. Of the 32 management directives reviewed for possible revision to reflect the elimination of the Executive Council and the realignment of the responsibilities of the Executive Director for Operations, Chief Financial Officer, and Chief Information Officer, issuance of two remains to be completed. One management directive is awaiting the Chairman's approval for issuance, as discussed in the paragraph in this section entitled "Review of Audit Follow-up System (OIG-00-A-05)." The other management directive is part of the management directive consolidation effort discussed in the paragraph in this section entitled "Review of the Agencywide Documents Access and Management System (OIG-02-A-12)." Issuance of these two remaining management directives, which is expected in March 2006, will complete agency actions on the OIG's recommendations from this audit.

Review of NRC's Quality Assurance Process for Official Documents (OIG-01-A-02)

February 23, 2001

The Office of the Inspector General recommended that NRC improve its quality assurance process for official documents by revising Management Directive and Handbook 3.57, *Correspondence Management*, to provide clear expectations for NRC staff to heighten awareness of the importance of information accuracy. Specifically, the Office of the Inspector General recommended that NRC establish the responsibilities of the document originator and concurrence chain reviewers with regard to accuracy of final products and to set expectations for document originators concerning fact-checking methods. Interim policy guidance on ensuring the technical accuracy and readability of NRC's documents and correspondence was issued to all NRC employees in May 2001. The revised management directive and handbook incorporating this policy and other needed updates are being finalized and are expected to be approved by the Executive Director for Operations for issuance in early FY 2006, which will complete agency actions on the Office of the Inspector General's recommendations from this audit.

Government Performance and Results Act: Review of the FY 1999 Performance Report (OIG-01-A-03)

February 23, 2001

The Office of the Inspector General recommended that NRC develop the management control procedures needed to produce valid and reliable performance data, including guidance on reporting unmet goals. Interim guidance for performance management and reporting performance information was issued in July 2001. In July 2002, a new Management Directive and Handbook 4.8, *Performance Measurements*, was issued for intra-agency review and comment. It was subsequently decided that performance measurement should be addressed in the broader context of budget and performance integration. Therefore, Management Directive 4.8 is being incorporated into a revised Management Directive and Handbook 4.7, *Planning, Budgeting, and Performance Management*. The revised Management Directive 4.7 will clarify roles and responsibilities in setting the agency's strategic direction, determining planned activities and resources, measuring and monitoring performance, and assessing performance. The revised Management Directive and Handbook 4.7 are scheduled to be issued for intra-agency review and comment in FY 2006. Issuance of this management directive and handbook will complete agency actions on the Office of the Inspector General's recommendations from this audit.

Review of the Agencywide Documents Access and Management System (OIG-02-A-12)

June 12, 2002

The Office of the Inspector General recommended that NRC finalize and issue its draft new management directive and handbook addressing the agency's systems development life-cycle management (SDLCM) methodology. In early FY 2003, NRC conducted a lessons-learned analysis to identify changes to improve the SDLCM methodology's effectiveness and usability. Feedback from this analysis resulted in major process revisions, which were documented in a draft new Management Directive 2.5, *Application Systems Life-Cycle Methodology*, and a draft new Handbook 2.5, *Systems Development and Life-Cycle Management Methodology*, which were expected to be issued by November 2004.

To address numerous comments on draft Management Directive and Handbook 2.5 from users, it has been consolidated with two existing management directives and associated handbooks–Management Directive and Handbook 2.1, *Information Technology Architecture*, and Management Directive and Handbook 2.2, *Capital Planning and Investment Control*. The result is a new draft directive, Management Directive 2.8, *Project Management Methodology*, which is accompanied by a Web-based manual. Pending intra-agency review and comment, the drafts are available on the internal NRC Web site. Issuance of Management Directive 2.8 and its manual are expected in March 2006 and will complete agency actions on the Office of the Inspector General's recommendations from this audit.

Review of Security at NRC Headquarters (OIG-02-A-14)

August 15, 2002

Due to the sensitive nature of the Office of the Inspector General's review and recommendations in this area, specific details are not furnished as part of this report. Completion of open recommendations has been delayed due to an increase in scope, including the acquisition of an adjacent lot as the primary entry and exit path to the NRC headquarters office complex, and due to the approvals required to make changes to the perimeter of the complex. Completion of agency actions on recommendations remaining open as of September 30, 2005, are expected to be completed by the end of 2005, which will complete agency actions on the Office of the Inspector General's recommendations from this audit.

Independent Evaluation of NRC's Information Security Program as Required by the Government Information Security Reform Act for FY 2002 (OIG-02-A-17)

September 11, 2002

Due to the sensitive nature of the Office of the Inspector General's review and recommendations in this area, specific details are not furnished as part of this report. As of September 30, 2005, completion of agency actions on this audit report requires the issuance of a new NRC management directive and its associated Web-based manual, which are discussed further in the paragraph in this section entitled "Review of the Agencywide Documents Access and Management System (OIG-02-A-12)." This agency action will be carried over to and tracked to completion via NRC's FY 2006 Plan of Action and Milestones required by the Federal Information Security Management Act.

Review of NRC's Handling and Marking of Sensitive Unclassified Information (OIG-03-A-01)

October 25, 2002

The Office of the Inspector General recommended that NRC revise the management directive and handbook governing the sensitive, unclassified information security program and mandate consistent use of markings. During FY 2004, the Executive Director for Operations sponsored an interoffice task force review of all internally and externally generated categories of sensitive, unclassified information, with the exception of safeguards information. The review focused on identifying where clarification may be appropriate in the requirements for marking, storage, access, transmission, reproduction, record keeping, and destruction of such information. The task force report and recommendations have been finalized and are being implemented. Implementation includes developing a revision of Management Directive and Handbook 12.6, *NRC Sensitive Unclassified Information Security Program*, which currently includes NRC policy and guidance for all types of sensitive unclassified information, including safeguards information.

The revised management directive and handbook will cover only sensitive unclassified non-safeguards information, and will simplify marking and other handling requirements. A new management directive and handbook will be issued to address safeguards information. The revised sensitive unclassified non-safeguards information management directive and handbook are expected to be issued in FY 2007. The new safeguards information management directive and handbook are expected to be issued in FY 2006. Until they are available, the new NRC policy for handling, marking, and protecting sensitive unclassified non-safeguards information is available on the internal Web site. Issuance of these revised documents will complete agency actions on the Office of the Inspector General's recommendations from this audit.

Use of Electronic Mail at NRC (OIG-03-A-11)

March 21, 2003

The Office of the Inspector General recommended that NRC revise Management Directive and Handbook 3.53, *NRC Records Management Program,* to include current information about the Agencywide Documents Access and Management System (ADAMS). NRC continues to work to consolidate Management Directive and Handbook 3.53, *NRC Records Management Program,* and Management Directive and Handbook 3.50, *Document Management,* and plans to issue one directive, to be entitled *NRC Records and Document Management Program.* This directive will have a two-part handbook, Part I of which will cover the NRC records management program, and Part II will address ADAMS document processing. As of the end of FY 2005, the draft was circulating for intra-agency review and comment, and the final is expected to be issued in FY 2006. Issuance of the revised management directive and handbook will complete agency actions on the recommendations from this audit.

NRC's Accountability for Special Nuclear Materials (OIG-03-A-15)

June 3, 2003

The Office of the Inspector General recommended several changes to strengthen NRC's oversight program for ensuring that licensees appropriately control and account for special nuclear material (SNM). The open recommendations, the agency actions required to address these recommendations, and projected completion dates for agency actions are as follows:

(1)   The Office of the Inspector General recommended that NRC conduct periodic inspections to verify that material licensees comply with material control and accounting (MC&A) requirements, including but not limited to visual inspections of licensees' SNM inventories and validation of report information. A comprehensive MC&A program review was undertaken by the NRC during 2004. Based on the results of this review and further staff analysis, in August 2005 the staff provided the Commission recommendations for MC&A program changes, which are still under consideration. Following receipt of the Commission's guidance, the staff will begin to implement any inspection program recommendations endorsed by the Commission. If the recommendations are endorsed by the Commission, NRC actions to address this recommendation may not be completed until FY 2007. In the interim, inspections of selected licensee responses to NRC Bulletin 2003-04 regarding inventories of source material and SNM tracked in the Nuclear Materials Management and Safeguards System (NMMSS) are being conducted in accordance with a temporary inspection instruction.

(2)   The Office of the Inspector General recommended that NRC staff report annually to the Commission on the effectiveness of NRC's inspection program for ensuring that licensees satisfactorily carry out their MC&A responsibilities. Performance measures were drafted and incorporated into the FY 2005 operating plan. Performance was evaluated monthly and quarterly during FY 2005, and MC&A inspection program highlights of interest were provided to the Commission during an annual program review briefing which was held in the second quarter of FY 2005. These activities will be continued until agreement is reached with the Office of the Inspector General that these NRC actions sufficiently address this recommendation.

(3)   The Office of the Inspector General recommended that NRC document the basis of the approach used to risk-inform NRC's oversight of MC&A activities for all types of materials licensees. A comprehensive MC&A program review was undertaken by the NRC during 2004. Based on the results of this review and further staff analysis, in August 2005 the staff provided the Commission recommendations for MC&A program changes, which are still under consideration. Following receipt of the Commission's guidance, the staff will begin to implement any MC&A oversight recommendations endorsed by the Commission, and will document the bases for any changes, including justification for recommendations that are risk-informed. If the recommendations are endorsed by the Commission, NRC actions to address this recommendation may not be completed until FY 2007.

(4)   The Office of the Inspector General recommended that NRC revise its regulations to require licensees authorized to possess SNM, and not currently required to do so, to conduct annual inventories and submit an annual Material Status Report or Physical Inventory Summary Report to NRC. A comprehensive MC&A program review was undertaken by the NRC during 2004. Based on the results of this review and further staff analysis, in August 2005 the staff provided the Commission recommendations for MC&A program changes, which are still under consideration. Following receipt of the Commission's guidance, the staff will begin to implement any rulemaking recommendations endorsed by the Commission. NRC actions to address this recommendation may not be completed until FY 2007.

(5)   The Office of the Inspector General recommended that NRC establish an independent system of accounting for SNM possessed by NRC and Agreement State licensees, and ensure that beginning balances are accurate based on NRC's physical verification of a statistical sample of the location and amounts of SNM held by the licensees, a review of a statistical sample of a licensee's records, or some combination thereof. Once this has been done, OIG has also recommended that NRC redirect its funding for NMMSS to the NRC licensee database, dissolve the current Department of Energy -NRC programmatic agreement for development and operation of NMMSS, and institute a new agreement relative to providing the Department of Energy with the information necessary to satisfy international SNM reporting obligations.

In lieu of abandoning NMMSS–to replace it with a new, independent NRC system of accounting for SNM–and the contractual relationship with the Department of Energy to maintain it, NRC is systematically addressing contributing causes of the concerns regarding the adequacy and integrity of the current NMMSS database. The NRC's activities in this regard include the NMMSS Rebaselining Project to facilitate the confirmation of licensee SNM holdings, direct NRC oversight of the NMMSS contractor's activities, and periodic coordination meetings with DOE to improve the effectiveness and efficiency of NMMSS operations and the NMMSS contractor's performance, among other efforts.

In May 2005, the staff provided recommendations to the Commission, including possible rulemaking changes that would enhance NMMSS accuracy for SNM. The Commission directed the staff to develop a rulemaking to incorporate the staff's recommendations, which was initiated in September 2005. Issuance of the final rule, which is expected in early 2008, will complete NRC actions to address this OIG recommendation. In the interim, the staff will continue its activities with DOE directed toward improving the availability and reliability of information in the NMMSS database. Additionally, the staff has completed a revision of Inspection Manual Chapter 2800, Material Inspection Program, and associated procedures to provide for confirmation of SNM inventories of reportable nuclear materials at licensee sites.

(6)   The Office of the Inspector General recommended that for any NRC funding to the Department of Energy directed toward meeting international reporting obligations, NRC should pursue all remedies available under agency policies and procedures for placement of and monitoring work with the Department of Energy. The NRC has pursued various actions with the Department of Energy to address the NMMSS contractor's performance issues. Closure of this recommendation requires confirmation that NRC is exercising all options available to encourage the Department of Energy 's compliance in submitting timely, monthly letter status reports as required by contract. It is expected that the Office of the Inspector General will complete its review of NRC's actions in early FY 2006.

Completion of the activities described above will complete agency actions on the Office of the Inspector General's recommendations from this audit.

Independent Evaluation of NRC's Implementation of the Federal Information Security Management Act for FY 2003 (OIG-03-A-22)

September 15, 2003

Due to the sensitive nature of the Office of the Inspector General's review and recommendations in this area, specific details are not furnished as part of this report. As of September 30, 2005, completion of agency actions on this audit report requires certification and accreditation of some systems and completion of contingency plan testing and documentation of findings and recommendations identified during the testing. These activities are expected to be completed by early 2006. These agency actions will be carried over to and tracked to completion via NRC's FY 2006 Plan of Action and Milestones required by the Federal Information Security Management Act.

Audit of the NRC's FY 2003 Financial Statements (OIG-04-A-03)

December 17, 2003

The Office of the Inspector General observed that NRC does not have a routine, timely, and disciplined process in place to monitor the adequacy of accounting information necessary to capitalize internal use software projects, and recommended that the Chief Financial Officer reassess the accounting activities being undertaken by agency personnel to ensure the completeness and propriety of accounting transactions. The Office of the Inspector General also recommended that the Chief Financial Officer be more proactive in monitoring and training project managers to instill discipline, thereby providing reliability of financial information.

In July 2004, the Chief Financial Officer issued revised procedures for monitoring approved software development projects to provide for more proactive monitoring and training of project managers and monitoring of accounting activities. During FY 2005, the Chief Financial Officer reassessed existing policies and procedures to improve the completeness and propriety of internal use software capitalization information. In addition, the Chief Financial Officer undertook a number of actions to educate NRC employees involved in software development projects about NRC's policies and procedures and their individual responsibilities and to ensure alignment of other agency directives with these policies and procedures. Closure of this recommendation and completion of agency actions on the recommendations from this audit requires Office of the Inspector General confirmation that NRC's actions have been sufficient. It is expected that the Office of the Inspector General will complete its review during its audit of NRC's FY 2005 financial statements.

Audit of NRC's Protection of Safeguards Information (OIG-04-A-04)

January 8, 2004

The Office of the Inspector General recommended that NRC finalize and issue the designation guidance document pertaining to an NRC category of sensitive unclassified information referred to as safeguards information and which was under development. The guide was completed on September 30, 2005, and was approved for issuance as an

NUREG. The NUREG will be published and made publically available before the end of 2005. The issuance of this NUREG will complete agency actions on the recommendations from this audit.

Review of NRC's Personnel Security Program (OIG-04-A-11)
March 25, 2004

The Office of the Inspector General recommended that in order to improve the likelihood that security clearances for summer interns are granted prior to or during their summer employment period, NRC should begin the hiring process for these interns one month earlier each year and should impose a deadline on them for returning their completed security package. The Office of the Inspector General also recommended that in accordance with Office of Personnel Management policy, NRC should inform the Office of Personnel Management when an intern terminates employment prior to completion of the Office of Personnel Management background investigation so that the investigation can be canceled.

The Office of Personnel Management has greatly improved on the lead time for processing security clearance investigations since this audit report was issued. In addition, the NRC is now using e-QIP (Electronic Questionnaires for Investigations Processing), which is an Internet-accessible tool for inputting and processing security questionnaire information that provides additional process efficiencies. Further, the Office of Personnel Management has agreed that ongoing investigations do not need to be terminated when a summer intern finishes their summer employment if the intern is expected to be re-employed at a later date. At the beginning of each fiscal year, the Office of Human Resources sends a request to NRC office directors asking them to initiate the hiring of summer interns. Beginning in FY 2006, this request will ask the office directors to identify interns from the previous summer who are not expected to return so that the Office of Personnel Management may be advised to terminate the investigations, if not yet completed, for these individuals. The request will also set an end-of-March deadline for submission of completed security packages by the interns. Closure of these recommendations and completion of agency actions on the recommendations from this audit requires the agency to provide confirmation to the Office of the Inspector General that the Office of Personnel Management has agreed to this approach.

Review of NRC's Reactor Operating Experience Task Force Report (OIG-04-A-13)

March 30, 2004

The Office of the Inspector General recommended that NRC revise its reactor operating experience program objectives to include measurable performance aspects, and establish an independent operating experience function and locate that function at the appropriate organizational level.

To incorporate measurable performance aspects into the program objectives, the NRC has expanded each program objective to include its supporting description and the attributes recommended in the Reactor Operating Experience Task Force report. This information has been incorporated into a new draft Management Directive and Handbook 8.7, *Reactor Operating Experience Program*, which was issued for interim use in December 2004. After the NRC staff has worked with the draft management directive and handbook for one year to receive feedback and identify if adjustments are needed, it will be finalized. It is expected that it will be ready for issuance in March 2006. In May 2005, a draft revised procedure was issued to describe in more detail the requirements, roles, and responsibilities of the Office of Nuclear Reactor Regulation as the designated lead for the NRC's reactor operating experience program. It is expected that the Office of Nuclear Reactor Regulation staff will use the draft revised procedure for about a year so that changes needed to improve the program and/or the process can be incorporated before it is issued in final form, which is expected to occur in mid-FY 2006. The NRC considers that the implementation of the organization responsibilities described in new draft Management Directive and Handbook 8.7 and in the revised Office of Nuclear Reactor Regulation procedure provides an appropriate interdependent organizational model for the reactor operating experience program, which when combined with the measurable performance aspects, sufficiently addresses Office of the Inspector General's recommendations. Issuance of the final management directive and handbook and final procedure will complete agency actions on the recommendations from this audit.

NRC's Implementation of Regulations Concerning Nondiscrimination Based on Handicap (OIG-04-A-14)

May 24, 2004

The Office of the Inspector General recommended that NRC revise Management Directive and Handbook 11.6, *Financial Assistance Program*, to identify and define the Office of Small Business and Civil Rights' role in accordance with 10 CFR Part 4, Subpart B, "Regulations Implementing Section 504 of the Rehabilitation Act of 1973, as Amended." The revised management directive and handbook are undergoing final review for approval, and are expected to be issued by the end of 2005, which will complete agency actions on the recommendations from this audit.

Review of NRC's Drug-Free Workplace Plan (OIG-04-A-15)

May 24, 2004

The Office of the Inspector General recommended that NRC revise the *NRC Drug-Free Workplace Plan* to include the clause on deferral of testing from the U.S. Department of Health and Human Services, *Model Plan for a Comprehensive Drug-Free Workplace Program*, which would allow drug testing of employees who were absent on the day of the drug test for up to 60 days after the test date. The Office of the Inspector General also recommended that NRC include instructions in the NRC plan that revisions must receive approval from Health and Human Services prior to implementation and obtain Health and Human Services approval of the 2004 version of the NRC plan prior to implementation.

The latest version of Revision 2 of the NRC plan, which was forwarded to Health and Human Services for review and approval on August 12, 2005, includes the deferral of testing clause and the instructions that revisions of the plan must receive approval from the Department of Health and Human Services prior to implementation. The Department of Health and Human Services provided comments on this version to NRC in early September 2005, and discussions are under way with Health and Human Services to ensure NRC's understanding of the comments. Another version of Revision 2 of the NRC plan, incorporating the Department of Health and Human Services 's latest comments, is expected to be provided to Health and Human Services for review and approval by the end of October 2005. The NRC will obtain Department of Health and Human Services approval prior to dissemination and implementation of Revision 2 of the NRC plan, which will complete agency actions on the recommendations from this audit.

AGREEMENT STATES

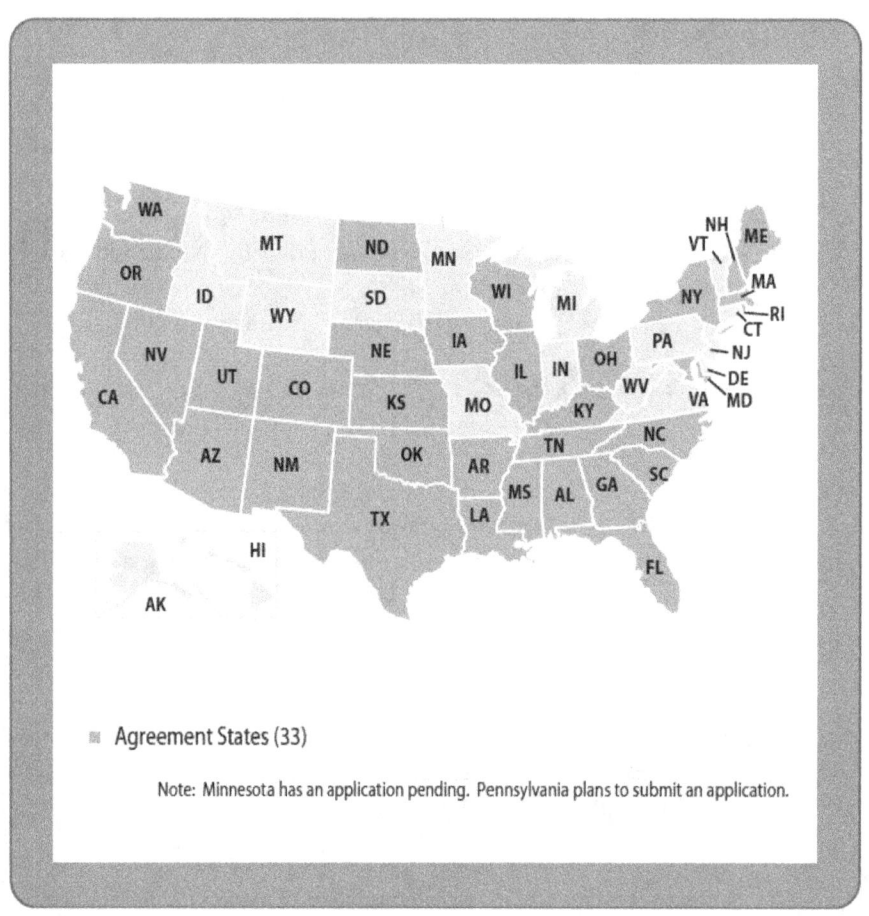

Agreement States (33)

Note: Minnesota has an application pending. Pennsylvania plans to submit an application.

Alabama	Illinois	Massachusetts	North Carolina	Tennessee
Arkansas	Iowa	Mississippi	North Dakota	Texas
Arizona	Kansas	Nebraska	Ohio	Utah
California	Kentucky	Nevada	Oklahoma	Washington
Colorado	Louisiana	New Hampshire	Oregon	Wisconsin
Florida	Maine	New Mexico	Rhode Island	
Georgia	Maryland	New York	South Carolina	

NRC ORGANIZATION CHART AS OF
SEPTEMBER 30, 2005

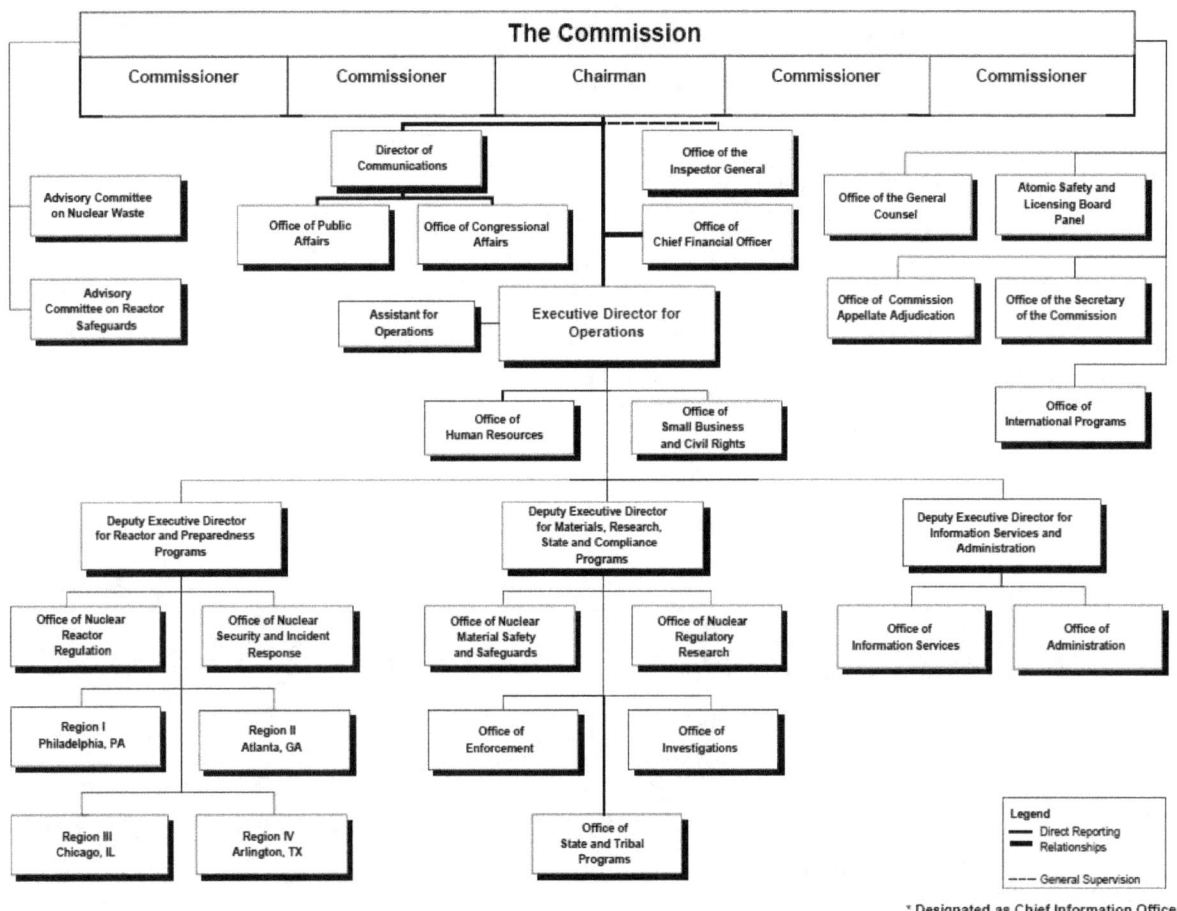

GLOSSARY OF ACRONYMS

ACRS	Advisory Committee on Reactor Safeguards
ADAMS	Agencywide Documents Access and Management System
AICPA	American Institute of Certified Public Accountants
AID	Aid for International Development
AO	abnormal occurrence
ASP	accident sequence precursor
CCR	Central Contractor Registration
CE	Combustion Engineering Owner's Group
CFO	Chief Financial Officer
CFO Act	Chief Financial Officers Act of 1990
CFR	Code of Federal Regulations
CIO	Chief Information Officer
CIOC	CIO Council
COLs	Combined Operating Licenses
CSRS	Civil Service Retirement System
CY	calendar year
DHS	Department of Homeland Security
DOE	Department of Energy
DOI	Department of Interior
DOL	Department of Labor
EDO	Executive Director for Operations
EFT	electronic funds transfer

E-Gov	electronic Government
EPA	Environment Protection Agency
E-QIP	Electronic Questionnaires for Investigations Processing
ESP	Early Site Permits
FACTS I	Federal Agencies' Centralized Trial Balance System
FAR	Federal Acquisition Regulation
FECA	Federal Employees Compensation Act
FEMA	Federal Emergency Management Agency
FERS	Federal Employees Retirement System
FFMIA	Federal Financial Management Improvement
FFS	Federal Financial System
FICA	Federal Insurance Compensation Act
FISMA	Federal Information Security Management Act
FPPS	Federal Personnel and Payroll System
FY	fiscal year
GAO	Government Accountability Office
GFE	Generic Fundamentals Examination
GFRS	Governmentwide Financial Reporting System
GPEA	Government Paperwork Elimination Act
GSA	General Services Administration
GSI	General Safety Issue
HLW	High-Level Waste
HSPD-12	Homeland Security Presidential Directive 12
IAEA	International Atomic Energy Agency
IG	Inspector General

IMPEP	Integrated Materials Performance Evaluation Program
Improvement Act	Federal Management Improvement Act of 1996
Integrity Act	Federal Managers' Financial Integrity Act of 1982
IOAA	Independent Offices Appropriation Act
IPAC	Intragovernment Payment and Collection
ISA	integrated safety analysis
IT	information technology
JFMIP	Joint Financial Management Information Program
LSN	Licensing Support Network
MC&A	material control and accounting
MOX	mixed-oxide fuel
MWe	Megawatts electric
NARA	National Archive and Records Administration
NBC	National Business Center
NFPA	National Fire Protection Association
NMED	Nuclear Materials Event Database
NMMSS	Nuclear Materials Management and Safeguards System
NMSS	Office of Nuclear Materials Safety and Safeguards
NRC	Nuclear Regulatory Commission
NRR	Office of Nuclear Reactor Regulation
NSIR	Office of Nuclear Security and Incident and Response
NUREG	Nuclear Regulatory Commission Regulation
NWF	Nuclear Waste Fund

OBRA-90	Omnibus Budget Reconciliation Act of 1990
OCFO	Office of the Chief Financial Officer
OEDO	Office of the Executive Director for Operations
OIG	Office of the Inspector General
OIS	Office of Information Services
OMB	Office of Management and Budget
OPM	Office of Personnel Management
OSART	Operational Safety Review Team
PAR	Performance and Accountability Report
PART	Program Assessment Rating Tool
PBPM	planning, budgeting, and performance management
PL	Public Law
PMM	Project Management Methodology
PRA	Probabilistic risk assessment
PRB	Petition Review Board
RASP	Risk Assessment Standardization Project
RES	Office of Nuclear Regulatory Research
RIRIP	Risk-Informed Regulation Implementation Plan
RLO	records liaison officer
RMG	records management guideline
ROP	reactor oversight process
SDLCM	system development life-cycle management
SDLCMM	system development life-cycle management methodology
SDP	Significance Determination Process
SECY	Office of the Secretary of the Commission

SFFAS	Statements of Federal Financial Accounting Standards
SFFAS Number 10	Accounting for Internal Use Software
SGI	Safeguards Information
SNM	special nuclear material
TI	temporary instruction
TSP	Thrift Savings Plan

1. "Nuclear reactor accidents" are defined in the NRC Severe Accident Policy Statement as those events that result in substantial damage to the reactor fuel, whether or not serious offsite consequences occur.

2. This measure is the number of new red inspection findings during the fiscal year plus the number of new red performance indicators during the fiscal year. Programmatic issues at multi-unit sites that result in red findings for each individual unit are considered separate conditions for purposes of reporting for this measure. A red performance indicator and a red inspection finding that are due to an issue with the same underlying causes are also considered separate conditions for purposes of reporting for this measure. Red inspection findings are included in the fiscal year in which the final significance determination was made. Red performance indicators are included in the fiscal year in which Reactor Oversight Process external Web page was updated to show the red indicator.

3. Significant Accident Sequence Precursor (ASP) events have a conditional core damage probability (CCDP) or ΔCDP of $\geq 1 \times 10^{-3}$. Such events have a $1/1000$ (10^{-3}) or greater probability of leading to a reactor accident involving core damage. An identical condition affecting more than one plant is counted as a single ASP event if a single accident initiator would have resulted in a single reactor accident. One event was identified in FY 2002 as having the potential of being a "significant" precursor. This precursor involved a reactor pressure vessel head degradation at Davis-Besse (see page 29 of last year's report). Preliminary Accident Sequence Precursor analysis shows Davis-Besse as a significant precursor. It will be final after the licensee comments. Based on the screening and engineering evaluation of FY 2002 and 2003 events, no other potentially "significant" precursors were identified. Therefore, the second performance measure was not exceeded for FY 2002 and 2003. For FY 2004 events occurring before June 1, 2004, screening and engineering evaluation of these events identified no potentially "significant" precursors.

4. This measure is the number of plants that have entered the Manual Chapter 0350 process, the multiple/repetitive degraded cornerstone column, or the unacceptable performance column during the fiscal year (i.e., were not in these columns or process the previous fiscal year). Data for this measure is obtained from the NRC external Web Action Matrix Summary page that provides a matrix

of the five columns with the plants listed within their applicable column and notes the plants in the Manual Chapter 0350 process. For reporting purposes, plants that are the subject of an approved deviation from the Action Matrix are included in the column or process in which they appear on the Web page.

5. Considering all indicators qualified for use in reporting.

6. Releases for which a 30-day report requirement under 10 CFR 20.2203(a)(3) is required.

7. With no event exceeding Abnormal Occurrence Criterion 1.B.1.

8. Performance targets have changed from FY 2000 to FY 2002 to reflect additional historical data. (Targets were as follows: FY 2000, 356; FY 2001, 350; FY 2002, 300; FY 2003, 300; and FY 2004, 300.)

9. Events of material entering the public domain in an uncontrolled manner are reported under 10 CFR 20.2201(a)(1)(i) and (ii). The NMED lists these events as reported by NRC licensees and, through the Agreement States, the Agreement State licensees. Data sources and verification: Events meeting this threshold could be reported to the NRC and/or Agreement States through a number of sources but primarily through licensee notifications. The Materials Inspection program is a key element in verifying the completeness and accuracy of licensee reports.

10. Performance targets have changed from FY 2000 to FY 2002 to reflect additional historical data. (Targets were as follows: FY 2000, 19; FY 2001, 40; FY 2002, 30; FY 2003, 30; and FY 2004, 30.)

11. Overexposures are those that exceed the dose limits specified in 10 CFR 20.2203(a)(2) as tracked in NMED. For fuel cycle activities, this extends to other hazardous materials used with, or produced from, licensed material, consistent with 10 CFR Part 70. Reportable chemical exposures are those that exceed license commitments. Such events would also include chemical exposures involving uranium recovery activities under the Uranium Mill Tailings Radiation Control Act. Multiple people may be affected by a single causal event. Data sources and verification: Events meeting this threshold could be reported to the NRC and/or Agreement States through a number of sources, but primarily through licensee notifications. The Materials Inspection program is a key element in verifying the completeness and accuracy of licensee reports. The IMPEP also verifies the accuracy of the event reports.

12. Medical events (misadministrations), as reported under 10 CFR Part 35, are tracked in NMED. Multiple patients may be affected by a single causal event. Data sources and verification: Events meeting this threshold could be reported to the NRC and/or Agreement States through a number of sources, but primarily through licensee notifications. The Materials Inspection program is a key element in verifying the completeness and accuracy of licensee reports.

13. This involves chemical releases from NRC-regulated activities under the Uranium Mill Tailings Radiation Control Act. Data sources and verification: Events meeting this threshold could be reported to the NRC and/or Agreement States through a number of sources, but primarily through licensee notifications. The Materials Inspection program is a key element in verifying the completeness and accuracy of licensee reports. Releases that cause impacts to the environment that cannot be mitigated within applicable regulatory limits using reasonably available methods are not readily defined. The expert judgement of NRC personnel and that of other agencies, such as the Environmental Protection Agency, are relied upon to make that determination. Events of this magnitude would result in a prompt and thorough investigation.

14. Defined as a disclosure that harms national security or public safety.

AVAILABILITY OF REFERENCE MATERIALS
IN NRC PUBLICATIONS

www.ingramcontent.com/pod-product-compliance
Lightning Source LLC
Chambersburg PA
CBHW081439170526
45166CB00008B/2255